Standards-Based Math

7-8

Written by
Alaska Hults

Editor: Collene Dobelmann
Illustrator: Corbin Hillam
Designer/Production: Moonhee Pak/Rosa Gandara
Cover Designer: Barbara Peterson
Art Director: Tom Cochrane
Project Director: Carolea Williams

Table of Contents

Introduction 4

Natural Numbers and Decimals

Natural Numbers and Decimals 5
Rational and Irrational Decimals 6
Rounding Decimals 7
Estimating Decimal Products
    and Quotients 8
Multiplying Decimals. 9
Dividing Decimals 10
Exponents. 11
Exponential Notation 12

Factors and Fractions

Divisibility. 13
Factors, Primes, and Composites 14
Prime Factorization 15
Least Common Multiple 16
Greatest Common Factor 17
Simplifying Fractions. 18
Functions 19
Comparing Fractions 20
Improper Fraction to Mixed Number. . . 21
Mixed Number to Improper Fraction. . . 22
Adding and Subtracting Fractions 23
Adding Mixed Numbers 24
Subtracting Mixed Numbers. 25
Factors and Fractions. 26
Fractions to Percents 27
Multiplying Fractions. 28
Multiplying Mixed Numbers. 29
Dividing Fractions 30
Dividing Mixed Numbers 31
Multistep Problems 32
Along the Way. 33
Altogether 34

Algebra

Variables. 35
Associative and Commutative
    Properties. 36
Distributive Property 37
Order of Operations 38
Inverse Operations. 39
Inverse Operations: Addition 40
Inverse Operations: Subtraction 41
Inverse Operations: Multiplication. 42
Inverse Operations: Division 43
Variables 2 44
Inequalities. 45
Graphing Inequalities 46

Geometry

Lines and Line Segments. 47
Angles. 48
Angle Measurement 49
Geometry Terms. 50
Polygons. 51
Similar and Congruent 52
Similar Triangles. 53
Similar Figures 54
Parts of a Circle 55
Geometric Solids 56
Spatial Relationships. 57
Angles, Circles, and Solids. 58
The X, Y Axis 59
Using a Cartesian Plane. 60
Reflections Across the X-Axis 61
Reflections Across the Y-Axis 62
Translations Along the X-Axis. 63
Translations Along the Y-Axis. 64
Oblique Translations 65
Transformations. 66
Perimeter 67
Area 68
Area of a Triangle and Trapezoid 69
Circumference 70

Area of Circles 71
Area and Circumference of Circles. 72
Points, Shapes, and Solids. 73

Ratio and Proportion

Ratios 74
Rates. 75
Proportions 76
Cross Multiplication. 77
Proportions 2 78
Unit Price 79
Ratio and Proportion 80

Data Analysis and Problem Solving

Percent. 81
Fractions, Decimals, and Percents. 82
More Fractions, Decimals,
    and Percents 83
Percent of a Number. 84
Data Analysis 85
Mean, Median, Mode, and Range 86
Reading Tables. 87
Reading Line Graphs. 88
Creating a Circle Graph. 89
Scatter Plots. 90
Reading Graphs 91
The Bottom Line 92

Integers

Defining Integers. 93
Absolute Value 94
Adding Integers. 95
Subtracting Integers 96
Exploring Linear Equations. 97
Positive and Negative 98

Rational Numbers

Terminating Decimals 99
Repeating Decimals. 100
Comparing Rational Numbers 101
The Density Property. 102
Scientific Notation. 103
Your Checkbook 104
Be Rational! 105

Probability

Combinations. 106
Permutations 107
Impossible and Certain 108
Calculating Chance 109

Answer Key 110

Introduction

Each book in the Power Practice™ series contains dozens of ready-to-use activity pages to provide students with skill practice. The fun activities can be used to supplement and enhance what you are already teaching in your classroom. Give an activity page to students as independent class work, or send the pages home as homework to reinforce skills taught in class. An answer key is included at the end of each book to provide verification of student responses.

Standards-Based Math 7–8 provides activities that will directly assist students in practicing basic skills and concepts. The structure of the book enhances student learning and enables them to meet the next challenge with confidence. Students will receive reinforcement in skills from the following math strands:
- Natural Numbers and Decimals
- Factors and Fractions
- Algebra
- Geometry
- Ratio and Proportion
- Data Analysis and Problem Solving
- Integers
- Rational Numbers
- Probability

Use these ready-to-go activities to "recharge" skill review and give students the power to succeed!

Name _____ Date _____

Natural Numbers and Decimals

NATURAL NUMBERS AND DECIMALS

> **Natural numbers** are the counting numbers. They are greater than zero and represent only whole quantities.
> **Decimals** represent some part of a whole. Example: 3.5 represents three wholes and one-half of a whole.

Identify each as a natural number or decimal.

① 15.0

② 98.6

③ 164

④ 74.$\overline{6}$

_____

_____

_____

_____

⑤ 9020.00

⑥ 182.75

⑦ 98.33

⑧ 334.433

_____

_____

_____

_____

Solve.

⑨
$$\begin{array}{r} 273 \\ \times\ 975 \\ \hline \end{array}$$

⑩
$$\begin{array}{r} 564 \\ \times\ 328 \\ \hline \end{array}$$

⑪
$$\begin{array}{r} 987 \\ \times\ 225 \\ \hline \end{array}$$

⑫
$$\begin{array}{r} 619 \\ \times\ 278 \\ \hline \end{array}$$

⑬
$$\begin{array}{r} 369 \\ \times\ 751 \\ \hline \end{array}$$

⑭
$$\begin{array}{r} 462 \\ \times\ 354 \\ \hline \end{array}$$

⑮
$$\begin{array}{r} 826 \\ \times\ 391 \\ \hline \end{array}$$

⑯
$$\begin{array}{r} 198 \\ \times\ 743 \\ \hline \end{array}$$

⑰
$$\begin{array}{r} 592 \\ \times\ 489 \\ \hline \end{array}$$

⑱
$$\begin{array}{r} 1295 \\ \times\ 336 \\ \hline \end{array}$$

⑲
$$\begin{array}{r} 3840 \\ \times\ 585 \\ \hline \end{array}$$

⑳
$$\begin{array}{r} 6067 \\ \times\ 241 \\ \hline \end{array}$$

Rational and Irrational Decimals

NATURAL NUMBERS AND DECIMALS

Decimals that are **rational numbers** can be turned into either repeating or terminating decimals.
  - 0.5 and 0.25 are terminating decimals.
  - $0.\overline{3}$, $0.1\overline{6}$, and $0.\overline{142857}$ are repeating decimals with a pattern that never ends.

Decimals that are **irrational numbers** go beyond the point that we can calculate them.
  - $\sqrt{7}$ = 2.6457513110645905016157536392604257102591830824501803 68 . . .
  - π = 3.14159265358979323846264338327950288419716939937510582097 . . .

Translate each fraction or square root into a decimal. Write each fraction from the fraction box beneath the appropriate heading. Find the irrational numbers and write them under the appropriate heading.

Fraction Box

$\frac{1}{2}$	$\frac{1}{3}$	$\sqrt{9}$	$\frac{1}{4}$	$\frac{1}{5}$	$\sqrt{25}$	$\frac{1}{6}$	$\frac{1}{7}$	π	$\frac{1}{8}$	$\frac{1}{9}$	$\frac{1}{10}$	$\sqrt{2}$	$\frac{2}{3}$	$\frac{3}{4}$	$\frac{2}{5}$
$\frac{5}{6}$	$\sqrt{5}$	$\frac{2}{7}$	$\frac{5}{7}$	$\sqrt{3}$	$\frac{3}{8}$	$\frac{2}{9}$	$\sqrt{6}$	$\frac{7}{9}$	$\frac{3}{10}$	$\sqrt{4}$	$\frac{3}{5}$	$\frac{6}{7}$	$\frac{5}{8}$	$\frac{4}{9}$	$\frac{8}{9}$

Terminating	Repeating

Irrational

Did you know? The philosopher Hippasus used geometric methods to prove that $\sqrt{2}$ is irrational. This so irritated the other mathematical philosophers that they threw him overboard. How's that for irrational?

Rounding Decimals

NATURAL NUMBERS AND DECIMALS

2	4	1	6	
Ones	Tenths	Hundredths	Thousandths	Ten Thousandths

Round 2.416 to the nearest tenth.
2.416 ⟶ 2.4
Round 2.416 to the nearest hundredth.
2.416 ⟶ 2.42

Round each number to the indicated place value.

1 Round 12.3456 to the nearest tenth _____

2 Round 3.0345 to the nearest thousandth _____

3 Round 7.789 to the nearest hundredth _____

4 Round 2.15672 to the nearest ten-thousandth _____

5 Round 3.45499 to the nearest hundredth _____

6 Round 9.012 to the nearest tenth _____

7 Round 3.6743 to the nearest thousandth _____

8 Round 315.697243 to the nearest thousandth _____

9 Round 654.145419 to the nearest hundredth _____

10 Round 0.012 to the nearest tenth _____

11 Round 1.629543 to the nearest thousandth _____

12 Round 98.9542 to the nearest tenth _____

13 Round 6.00003 to the nearest ten-thousandth _____

14 Round 3.9999 to the nearest thousandth _____

Standards-Based Math • 7–8 © 2004 Creative Teaching Press

Name _____ Date _____

Estimating Decimal Products and Quotients

NATURAL NUMBERS AND DECIMALS

```
    3. 2 9 2                                                              3
  × 4. 7 0 8   Estimate the answer. Round to the nearest whole number.  × 5
```

Round each decimal to the nearest whole number. Write the estimated product or quotient. Do the work in your head, not on paper.

① 1.2×3.15

② 7.8×50.2

③ 5.4×8.2

④ 9.5×7.11

⑤ 89.7×9.6

⑥ 1.5×8.24

⑦ 7×40.2

⑧ 8×79.6

⑨ 54.5×4.86

⑩ $34.32 \div 7.4$

⑪ $23.6 \div 6.3$

⑫ $55.8 \div 8.24$

⑬ $299.7 \div 2.34$

⑭ $100.46 \div 24.8$

⑮ $74.86 \div 2.91$

⑯ $399.5 \div 39.7$

⑰ $71.86 \div 7.8$

⑱ $36.4 \div 11.9$

⑲ $11.672 \div 4.4$

⑳ $559.9 \div 56.2$

㉑ $42.2 \div 14.399$

Standards-Based Math • 7–8 © 2004 Creative Teaching Press

Multiplying Decimals

Natural Numbers and Decimals

$24.56 \times 3.123 = 76.70088$
  2   3   5

$0.00003 \times 0.0001 = 0.000000003$
  5   4   9

$24.56 \times 3.123 = 76.70088$ (5 decimal places total)
$0.00003 \times 0.0001 = 0.000000003$ (9 decimal places total)

Find the product.

1 $4.5 \times 10 =$

2 $5.244 \times 0.921 =$

3 $4.05 \times 3 =$

4 $60 \times 0.12 =$

5 $0.0335 \times 0.05 =$

6 $102 \times 0.921 =$

7 $5 \times 3.221 =$

8 $0.02 \times 0.12 =$

9 $12.4 \times 3.4 \times 2.7 =$

10 $100 \times 0.001 =$

11 $24.02 \times 33.2 \times 4.1 =$

12 $1.1 \times 2.2 \times 3.3 \times 4.4 =$

13 $10 \times 5.2 \times 100 \times 5.45 =$

14 $100 \times 0.01 \times 1000 \times 0.02000 \times 0.3 =$

15 $6.7 \times 100 \times 2.3 \times 0.4 =$

16 $4 \times 0.001 \times 250 \times 100 \times 0.2000 =$

17 $0.0900 \times 10 \times 0.003 \times 100 =$

18 $8.4 \times 0.35 \times 18.2 =$

19 $4.4 \times 10.10 \times 25 \times 0.0025 =$

20 $10 \times 0.1 \times 100 \times 0.0100 \times 10 \times 0.05 =$

Name _____   Date _____

Dividing Decimals

NATURAL NUMBERS AND DECIMALS

$$40 \div 0.02 = 0.02\overline{)40.00} = 2\overline{)4000}^{\,2000}$$

Solve.

1 $4.6 \div 2.3 =$

2 $7.5 \div 0.25 =$

3 $110 \div 1.1 =$

4 $60 \div 0.12 =$

5 $810 \div 0.09 =$

6 $310.5 \div 4.5 =$

7 $5 \div 2.5 =$

8 $40.32 \div 1.6 =$

9 $128.57544 \div 2.367 =$

10 $100 \div 0.001 =$

11 $150 \div 0.25 =$

12 $56.2 \div 0.02 =$

13 $66.0543 \div 0.001 =$

14 $10000 \div 0.05 =$

15 $82.02 \div 0.02 =$

16 $30 \div 0.006 =$

17 $0.8 \div 0.025 =$

18 $25 \div 5.5 =$

19 $0.6 \div 0.001 =$

20 $56.97 \div 3.6 =$

Standards-Based Math • 7–8 © 2004 Creative Teaching Press

Name _____   Date _____

Exponents

NATURAL NUMBERS AND DECIMALS

An **exponent** next to a number is a short way of writing a multiplication problem in which a number is multiplied by itself. The exponent tells how many times the number appears in the multiplication problem.

$$2 \times 2 = 2^2$$

$$2 \times 2 \times 2 \times 2 = 2^4$$

We read a number with an exponent as (number) to the (exponent) power.

2^4 = two to the fourth power 2^2 = two to the second power

Write out the multiplication problem represented by each number and exponent. Solve.

1 5^3

2 2^4

3 3^4

4 6^2

5 8^3

6 9^2

7 5^4

8 7^3

9 4^3

10 10^3

11 12^2

12 11^3

13 1^{12}

14 4^4

15 7^5

16 5^5

17 2^8

18 6^3

19 9^4

20 3^5

21 6^4

Exponential Notation

Natural Numbers and Decimals

> The number 10 with an exponent is special. The exponent tells you exactly how many zeros come after the decimal point.
>
> $$10^6 = 10 \cdot 10 \cdot 10 \cdot 10 \cdot 10 \cdot 10 = 1,000,000$$

Write out the number each power of ten represents.

1 10^3 _____

2 10^8 _____

3 10^7 _____

> Scientific notation uses powers of ten to write long numbers in a short way. The power of ten tells you how many places to move the decimal point and in which direction. A positive exponent means move the decimal point to the right. A negative exponent means move the decimal to the left.
>
> $$8,340,000,000,000 = 8.34 \cdot 10^{12}$$
> $$0.00000000005 = 5 \cdot 10^{-11}$$

Write the number in standard form.

4 $7.2 \cdot 10^5$

_____

5 $6 \cdot 10^7$

_____

6 $8.7 \cdot 10^{-4}$

_____

7 $9 \cdot 10^9$

_____

8 $5.2 \cdot 10^{-8}$

_____

9 $1.204 \cdot 10^{-6}$

_____

10 $6 \cdot 10^{-7}$

_____

11 $4.2 \cdot 10^{12}$

_____

12 $1.11 \cdot 10^{10}$

_____

13 $3.879 \cdot 10^9$

_____

14 $6.8 \cdot 10^{-5}$

_____

15 $9.8 \cdot 10^9$

_____

16 $9.6 \cdot 10^{14}$

_____

17 $9.873 \cdot 10^{-8}$

_____

18 $9 \cdot 10^{-5}$

_____

Standards-Based Math • 7–8 © 2004 Creative Teaching Press

Name _____ Date _____

Divisibility
FACTORS AND FRACTIONS

A natural number is divisible by . . .

2 if the number is even.
3 if the sum of the digits is divisible by three.
4 if the number formed by its last two digits is divisible by 4.
5 if the last digit is a 5 or a 0.
6 if the number is divisible by both 2 and 3.
8 if the number formed by the last three digits is divisible by 8.
9 if the sum of the digits is divisible by 9.
10 if the last digit is a 0.

Example:
Is 98,536 divisible by 2, 3, 4, 5, 6, 8, 9, and 10?

2 Yes, because it is even.
3 No, because 9 + 8 + 5 + 3 + 6 = 31, which is not divisible by 3.
4 Yes, because 36 is divisible by 4.
5 No, because it doesn't end in a 5 or a 0.
6 No, because it is only divisible by 2 and not by 3.
8 Yes, because 536 is divisible by 8 (536 ÷ 8 = 67).
9 No, because 9 + 8 + 5 + 3 + 6 = 31 which is not divisible by 9.
10 No, because it doesn't end in a 0.

Determine if the following numbers are divisible by 2, 3, 4, 5, 6, 8, 9, and 10. Use the back of the paper if you need more room. The first one is done for you.

1 1,008

Yes: 2, 3, 4, 6, 8, 9
No: 5,10

2 3,024

Yes: _____
No: _____

3 920

Yes: _____
No: _____

4 518,400

Yes: _____
No: _____

5 186,624

Yes: _____
No: _____

6 319,599

Yes: _____
No: _____

7 4,520

Yes: _____
No: _____

8 92,801

Yes: _____
No: _____

9 4,904

Yes: _____
No: _____

10 21,866

Yes: _____
No: _____

11 3,044

Yes: _____
No: _____

12 11,985

Yes: _____
No: _____

13 76,767

Yes: _____
No: _____

14 61,854

Yes: _____
No: _____

15 71,006

Yes: _____
No: _____

16 47,422

Yes: _____
No: _____

17 21,089

Yes: _____
No: _____

18 19,941

Yes: _____
No: _____

Name _____ Date _____

Factors, Primes, and Composites

FACTORS AND FRACTIONS

A **factor** is a whole number that is multiplied by another whole number to equal a product.
Factors of 24: 1, 2, 3, 4, 6, 8, 12, and 24.

Prime: A natural number that has exactly two factors, the number itself and 1. Examples: 2, 3, and 5
Composite number: A natural number that has three or more factors. Examples: 4, 6, 8, and 9.
1 is neither prime nor composite.

Identify each number as prime or composite. If the number is composite, give at least one factor of the number that is not one or the number.

1 61 _____

2 16 _____

3 39 _____

4 42 _____

5 13 _____

6 12 _____

List the factors. Circle any factors that are also prime numbers. Use the back of the paper if more room is needed.

7 42

8 77

9 210

10 126

11 50

12 200

13 143

14 36

15 164

16 122

17 26

18 64

19 360

20 234

21 48

Standards-Based Math • 7–8 © 2004 Creative Teaching Press

Prime Factorization

Factors and Fractions

Name _____ **Date** _____

> **Prime Factorization:** Expressing a number as a product of primes.
>
>
>
>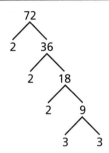
>
> Therefore, 50 = 5 × 5 × 2   Therefore, 26 = 2 × 13   Therefore, 72 = 2 × 2 × 2 × 3 × 3

Find the prime factorization or state that the number is prime. Show your work on a separate peice of paper.

1 66

2 325

3 23

4 451

5 494

6 29

7 674

8 99

9 225

10 369

11 184

12 208

13 493

14 148

15 293

16 132

17 75

18 128

19 529

20 48

21 68

Name _____ Date _____

Least Common Multiple

FACTORS AND FRACTIONS

A **least common multiple** is the multiple a pair or group of numbers have in common with the least value.

4: 4, 8, 12, 16, 20, <u>24,</u> 28, 32, 36, 40, 44, 48, 52, 56, 60
6: 6, 12, 18, <u>24,</u> 30, 36, 42, 48, 54, 60
8: 8, 16, <u>24,</u> 32, 40, 48, 56, 64

There is only one least common multiple for any group of numbers.

Find the least common multiple for each group of numbers.

1 5, 8:

2 7, 5:

3 3, 8:

4 9, 12, 6:

5 15, 10, 20:

6 12, 16:

7 9, 3:

8 4, 12:

9 4, 6, 8:

10 2, 6, 3:

11 8, 9, 12:

12 3, 7, 4:

Standards-Based Math • 7-8 © 2004 Creative Teaching Press

Greatest Common Factor

FACTORS AND FRACTIONS

A **greatest common factor** is the factor a pair or group of numbers have in common with the greatest value.

4: 4, 2, 1
6: 6, 3, 2, 1
8: 8, 4, 2, 1

36: 36, 18, 12, 9, 6, 4, 3, 2, 1
48: 48, 24, 16, 12, 8, 6, 4, 3, 2, 1

There is only one greatest common factor for any group of numbers.

Find the greatest common factor of each group of numbers.

1 24, 36:

2 36, 9:

3 21, 28:

4 32, 40, 16:

5 45, 15, 60:

6 12, 16:

7 15, 20:

8 93, 62:

9 52, 104:

10 117, 156:

Simplifying Fractions

Factors and Fractions

> To convert a fraction to its simplest terms, divide the numerator and denominator by their greatest common factor.
>
> $$\frac{21}{28} \div \frac{7}{7} = \frac{3}{4}$$

Reduce to simplest terms.

1 $\dfrac{6}{63} =$ **2** $\dfrac{36}{60} =$ **3** $\dfrac{8}{20} =$ **4** $\dfrac{18}{24} =$ **5** $\dfrac{24}{32} =$

6 $\dfrac{12}{16} =$ **7** $\dfrac{20}{60} =$ **8** $\dfrac{32}{48} =$ **9** $\dfrac{9}{54} =$ **10** $\dfrac{10}{16} =$

11 $\dfrac{50}{75} =$ **12** $\dfrac{18}{32} =$ **13** $\dfrac{31}{93} =$ **14** $\dfrac{11}{88} =$ **15** $\dfrac{32}{96} =$

16 $\dfrac{7}{35} =$ **17** $\dfrac{62}{93} =$ **18** $\dfrac{45}{99} =$ **19** $\dfrac{75}{225} =$ **20** $\dfrac{50}{500} =$

Standards-Based Math • 7–8 © 2004 Creative Teaching Press

Functions

Factors and Fractions

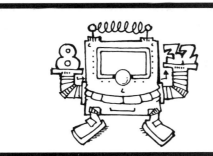

8 goes in, 32 comes out.
4 goes in, 16 comes out.
What is the rule?
× 4

Describe the rule. The first one is done for you.

1 4 goes in, 20 comes out.
6 goes in, 30 comes out. ___× 5___

2 14 goes in, 28 comes out.
9 goes in, 23 comes out. _____

3 12 goes in, 4 comes out.
39 goes in, 13 comes out. _____

4 12 goes in, 36 comes out.
9 goes in, 27 comes out. _____

5 31 goes in, 24 comes out.
65 goes in, 58 comes out. _____

6 42 goes in, 7 comes out.
84 goes in, 14 comes out. _____

7 9 goes in, 108 comes out.
12 goes in, 144 comes out. _____

8 16 goes in, 14 comes out.
87 goes in, 85 comes out. _____

9 76 goes in, 84 comes out.
25 goes in, 33 comes out. _____

10 14 goes in, 84 comes out.
20 goes in, 120 comes out. _____

11 45 goes in, 76 comes out.
26 goes in, 57 comes out. _____

12 112 goes in, 14 comes out.
200 goes in, 25 comes out. _____

13 16 goes in, 144 comes out.
70 goes in, 630 comes out. _____

14 1,456 goes in, 1,450 comes out.
23,899 goes in, 23,893 comes out. _____

15 8 goes in, 4 comes out.
18 goes in, 9 come out. _____

16 $\frac{7}{8}$ goes in, $\frac{5}{8}$ comes out.

$\frac{1}{2}$ goes in, $\frac{1}{4}$ comes out. _____

17 $\frac{1}{4}$ goes in, $\frac{3}{4}$ comes out.

$\frac{5}{8}$ goes in, $\frac{9}{8}$ comes out. _____

18 1 goes in, $\frac{1}{4}$ comes out.

4 goes in, 1 comes out. _____

Comparing Fractions

FACTORS AND FRACTIONS

Denominator: The number of equal sections of the whole. Also, the bottom number of the fraction.
Numerator: The number of equal sections that the fraction represents. Also, the top number of the fraction.

$$\frac{8}{24} \quad \frac{1}{3} \qquad \frac{1}{8} \quad \frac{3}{24}$$

$$\frac{1}{3} > \frac{1}{8}$$

Compare the fractions. Write the symbol that makes the statement correct.

1 $\frac{1}{8} \bigcirc \frac{1}{5}$

2 $\frac{2}{3} \bigcirc \frac{3}{5}$

3 $\frac{1}{4} \bigcirc \frac{1}{5}$

4 $\frac{2}{4} \bigcirc \frac{1}{2}$

5 $\frac{8}{9} \bigcirc \frac{9}{8}$

6 $\frac{1}{6} \bigcirc \frac{4}{25}$

7 $\frac{1}{7} \bigcirc \frac{71}{500}$

8 $\frac{3}{25} \bigcirc \frac{1}{8}$

9 $\frac{1}{5} \bigcirc \frac{5}{25}$

10 $\frac{6}{7} \bigcirc \frac{428,571}{500,000}$

11 $\frac{75}{150} \bigcirc \frac{1}{2}$

12 $\frac{12}{24} \bigcirc \frac{36}{56}$

13 $\frac{5}{6} \bigcirc \frac{30}{38}$

14 $\frac{48}{72} \bigcirc \frac{2}{3}$

15 $\frac{7}{8} \bigcirc \frac{49}{64}$

16 $\frac{22}{44} \bigcirc \frac{43}{88}$

17 $\frac{52}{63} \bigcirc \frac{8}{9}$

18 $\frac{1}{4} \bigcirc \frac{198}{780}$

19 $\frac{3}{4} \bigcirc \frac{252}{336}$

20 $\frac{11}{12} \bigcirc \frac{115}{144}$

Standards-Based Math • 7–8 © 2004 Creative Teaching Press

Improper Fraction to Mixed Number

FACTORS AND FRACTIONS

Divide the numerator by the denominator. Place the remainder over the denominator. Simplify.

$$\frac{26}{8}$$

$$8\overline{)26}\,^{3\,R2}$$

$$\frac{26}{8} = 3\frac{2}{8} = 3\frac{1}{4}$$

Simplify.

1 $\dfrac{10}{6} =$

2 $\dfrac{30}{8} =$

3 $\dfrac{28}{9} =$

4 $\dfrac{17}{7} =$

5 $\dfrac{13}{2} =$

6 $\dfrac{34}{5} =$

7 $\dfrac{39}{11} =$

8 $\dfrac{14}{3} =$

9 $\dfrac{68}{12} =$

10 $\dfrac{30}{19} =$

11 $\dfrac{127}{19} =$

12 $\dfrac{23}{4} =$

13 $\dfrac{44}{10} =$

14 $\dfrac{13}{6} =$

15 $\dfrac{48}{10} =$

16 $\dfrac{49}{11} =$

17 $\dfrac{29}{8} =$

18 $\dfrac{31}{6} =$

19 $\dfrac{98}{12} =$

20 $\dfrac{168}{15} =$

21 $\dfrac{215}{25} =$

22 $\dfrac{444}{6} =$

23 $\dfrac{77}{15} =$

24 $\dfrac{252}{10} =$

Name _____ Date _____

Mixed Number to Improper Fraction

Factors and Fractions

Multiply the denominator by the whole number. Add the product to the numerator. Place the sum over the denominator.

$$4\frac{3}{5}$$

$$5 \times 4 = 20 + 3 = 23$$

$$4\frac{3}{5} = \frac{23}{5}$$

Convert to an improper fraction.

1 $7\frac{5}{7} =$

2 $4\frac{1}{11} =$

3 $11\frac{2}{5} =$

4 $8\frac{2}{9} =$

5 $6\frac{3}{10} =$

6 $8\frac{2}{8} =$

7 $12\frac{5}{6} =$

8 $12\frac{3}{6} =$

9 $8\frac{6}{9} =$

10 $9\frac{1}{8} =$

11 $6\frac{4}{12} =$

12 $12\frac{4}{5} =$

13 $8\frac{3}{8} =$

14 $9\frac{1}{6} =$

15 $9\frac{1}{5} =$

16 $11\frac{1}{5} =$

17 $7\frac{2}{11} =$

18 $10\frac{5}{6} =$

19 $3\frac{7}{8} =$

20 $12\frac{2}{3} =$

21 $9\frac{3}{6} =$

22 $14\frac{1}{2} =$

23 $20\frac{3}{4} =$

24 $15\frac{7}{8} =$

Standards-Based Math • 7–8 © 2004 Creative Teaching Press

Name _____ Date _____

Adding and Subtracting Fractions

FACTORS AND FRACTIONS

Use the least common multiple to put fractions in the same terms. Add or subtract numerators.
Reduce to simplest terms using the greatest common factor.

$$\frac{1}{3}+\frac{1}{6}=\frac{2}{6}+\frac{1}{6}=\frac{3}{6}=\frac{1}{2} \qquad \frac{2}{3}-\frac{1}{6}=\frac{4}{6}-\frac{1}{6}=\frac{3}{6}=\frac{1}{2}$$

Solve.

1 $\dfrac{2}{3} + \dfrac{1}{4} =$

2 $\dfrac{2}{3} - \dfrac{2}{4} =$

3 $\dfrac{31}{72} - \dfrac{3}{8} =$

4 $\dfrac{7}{11} - \dfrac{2}{11} =$

5 $\dfrac{10}{15} + \dfrac{2}{15} =$

6 $\dfrac{11}{18} + \dfrac{5}{54} =$

7 $\dfrac{1}{6} + \dfrac{3}{4} =$

8 $\dfrac{5}{6} - \dfrac{3}{4} =$

9 $\dfrac{2}{3} - \dfrac{8}{27} =$

10 $\dfrac{6}{11} - \dfrac{4}{22} =$

11 $\dfrac{13}{24} + \dfrac{5}{12} =$

12 $\dfrac{5}{8} - \dfrac{1}{3} =$

13 $\dfrac{4}{15} + \dfrac{3}{5} =$

14 $\dfrac{3}{5} - \dfrac{4}{15} =$

15 $\dfrac{1}{5} - \dfrac{1}{6} =$

16 $\dfrac{7}{10} - \dfrac{3}{15} =$

17 $\dfrac{9}{12} + \dfrac{1}{8} =$

18 $\dfrac{7}{8} - \dfrac{8}{10} =$

19 $\dfrac{3}{8} + \dfrac{1}{24} =$

20 $\dfrac{3}{8} + \dfrac{2}{3} =$

21 $\dfrac{2}{3} + \dfrac{1}{6} + \dfrac{1}{12} =$

22 $\dfrac{2}{15} - \dfrac{1}{30} - \dfrac{1}{45} =$

23 $\dfrac{1}{2} + \dfrac{3}{8} + \dfrac{1}{12} =$

24 $\dfrac{1}{2} - \dfrac{1}{3} - \dfrac{1}{12} =$

Adding Mixed Numbers

Factors and Fractions

$$3\frac{1}{2} + 3\frac{1}{3} = \frac{7}{2} + \frac{7}{3} = \frac{21}{6} + \frac{14}{6} = \frac{35}{6} = 5\frac{5}{6}$$

Solve.

1 $4\frac{1}{8} + 2\frac{3}{4} =$

2 $5\frac{2}{6} + 3\frac{1}{9} =$

3 $7\frac{4}{5} + 8\frac{2}{15} =$

4 $3\frac{1}{9} + 3\frac{3}{4} =$

5 $6\frac{5}{8} + 9\frac{1}{12} =$

6 $8\frac{2}{5} + 3\frac{2}{3} =$

7 $7\frac{1}{4} + 6\frac{2}{5} =$

8 $9\frac{1}{12} + 9\frac{1}{3} =$

9 $14\frac{2}{3} + 12\frac{1}{2} =$

10 $4\frac{7}{8} + 2\frac{9}{10} =$

11 $7\frac{4}{7} + 5\frac{5}{7} =$

12 $5\frac{4}{9} + 3\frac{2}{3} =$

13 $7\frac{12}{13} + 5\frac{1}{10} =$

14 $3\frac{1}{12} + 6\frac{4}{9} =$

15 $3\frac{5}{9} + 8\frac{7}{16} =$

16 $5\frac{4}{5} + 7\frac{3}{8} =$

17 $5\frac{4}{9} + 3\frac{3}{7} =$

18 $6\frac{1}{5} + 9\frac{4}{7} =$

Standards-Based Math • 7–8 © 2004 Creative Teaching Press

Subtracting Mixed Numbers

FACTORS AND FRACTIONS

$$3\frac{1}{2} - 2\frac{3}{4} = \frac{7}{2} - \frac{11}{4} = \frac{14}{4} - \frac{11}{4} = \frac{3}{4}$$

Solve.

1 $5\frac{1}{8} - 2\frac{3}{4} =$

2 $9\frac{4}{15} - 6\frac{1}{5} =$

3 $5\frac{4}{9} - 3\frac{1}{3} =$

4 $7\frac{3}{8} - 5\frac{7}{24} =$

5 $8\frac{7}{16} - 3\frac{5}{24} =$

6 $6\frac{4}{9} - 3\frac{1}{12} =$

7 $7\frac{9}{10} - 5\frac{4}{10} =$

8 $5\frac{4}{9} - 2\frac{1}{6} =$

9 $7\frac{4}{7} - 1\frac{5}{14} =$

10 $4\frac{7}{8} - 3\frac{1}{2} =$

11 $14\frac{2}{3} - 12\frac{1}{2} =$

12 $9\frac{1}{3} - 4\frac{1}{6} =$

13 $7\frac{3}{4} - 6\frac{2}{5} =$

14 $8\frac{2}{5} - 5\frac{7}{20} =$

15 $9\frac{5}{12} - 4\frac{2}{9} =$

16 $3\frac{3}{4} - 1\frac{1}{8} =$

17 $8\frac{2}{15} - 4\frac{4}{45} =$

18 $5\frac{1}{3} - 4\frac{1}{3} =$

Factors and Fractions

FACTORS AND FRACTIONS

Respond.

1 Circle the irrational number: $\sqrt{4}$, $\sqrt{6}$, $\sqrt{25}$

2 Circle the repeating decimal: 0.25   0.$\overline{55}$ 0.55

3 Round to the nearest hundredth: 2.15672 _____

4 Estimate the product: 1.548782 × 8.246791 _____

5 Estimate the quotient: 23.64918 ÷ 6.37128 _____

Solve.

6 12.4 × 3.4 × 2.7

7 4.6 ÷ 2.3

8 5 × 3.221

9 10 ÷ 0.006

10 60 × 0.12

11 6^3

12 8.7×10^{-4}

13 $4\frac{7}{8} \times 2\frac{9}{10}$

14 $5\frac{1}{8} - 2\frac{3}{4}$

Standards-Based Math • 7–8 © 2004 Creative Teaching Press

Fractions to Percents

FACTORS AND FRACTIONS

$$\frac{1}{6} = 0.16\overline{6}$$

$$0.166 \approx 0.17$$
$$0.17 \times 100 = 17\%$$

$$\text{So, } \frac{1}{6} = 0.166 \approx 17\%$$

Convert the fractions to percents.

1 $\frac{2}{6} = 0.333 \approx$ _____ $\times 100 =$

2 $\frac{5}{18} = 0.2777 \approx$ _____ $\times 100 =$

3 $\frac{4}{12} = 0.333 \approx$ _____ $\times 100 =$

4 $\frac{5}{9} = 0.555 \approx$ _____ $\times 100 =$

5 $\frac{4}{21} = 0.190476 \approx$ _____ $\times 100 =$

6 $\frac{5}{24} = 0.208333 \approx$ _____ $\times 100 =$

7 $\frac{7}{27} = 0.259259 \approx$ _____ $\times 100 =$

8 $\frac{4}{33} = 0.1212 \approx$ _____ $\times 100 =$

9 $\frac{8}{24} = 0.333 \approx$ _____ $\times 100 =$

10 $\frac{5}{54} = 0.0925925 \approx$ _____ $\times 100 =$

11 $\frac{14}{20} =$

12 $\frac{9}{50} =$

13 $\frac{63}{70} =$

14 $\frac{45}{50} =$

15 $\frac{52}{80} =$

16 $\frac{31}{42} =$

17 $\frac{32}{60} =$

18 $\frac{12}{36} =$

19 $\frac{6}{14} =$

20 $\frac{56}{90} =$

Multiplying Fractions

FACTORS AND FRACTIONS

Simplify fractions, if needed.
Multiply the numerators. Multiply the denominators.
Simplify again, if needed.

$$\frac{2}{10} \times \frac{6}{9} = \frac{1}{5} \times \frac{2}{3} = \frac{1 \times 2}{5 \times 3} = \frac{2}{15}$$

Solve.

1 $\dfrac{3}{9} \times \dfrac{3}{7} =$

2 $\dfrac{3}{4} \times \dfrac{7}{8} =$

3 $\dfrac{2}{5} \times \dfrac{1}{3} =$

4 $\dfrac{6}{8} \times \dfrac{1}{3} =$

5 $\dfrac{4}{6} \times \dfrac{5}{9} =$

6 $\dfrac{5}{7} \times \dfrac{1}{5} =$

7 $\dfrac{3}{4} \times \dfrac{2}{4} =$

8 $\dfrac{4}{7} \times \dfrac{2}{4} =$

9 $\dfrac{1}{3} \times \dfrac{4}{7} =$

10 $\dfrac{8}{9} \times \dfrac{4}{8} =$

11 $\dfrac{1}{3} \times \dfrac{6}{7} =$

12 $\dfrac{3}{4} \times \dfrac{1}{5} =$

13 $\dfrac{1}{7} \times \dfrac{5}{7} =$

14 $\dfrac{3}{5} \times \dfrac{1}{5} =$

15 $\dfrac{5}{8} \times \dfrac{3}{7} =$

16 $\dfrac{4}{8} \times \dfrac{1}{7} =$

17 $\dfrac{1}{3} \times \dfrac{3}{8} =$

18 $\dfrac{2}{7} \times \dfrac{7}{8} =$

19 $\dfrac{1}{3} \times \dfrac{3}{5} \times \dfrac{5}{7} =$

20 $\dfrac{3}{4} \times \dfrac{4}{5} \times \dfrac{1}{3} =$

21 $\dfrac{5}{2} \times \dfrac{2}{3} \times \dfrac{3}{10} =$

22 $\dfrac{5}{12} \times \dfrac{6}{7} \times \dfrac{2}{3} =$

23 $\dfrac{4}{5} \times \dfrac{2}{7} \times \dfrac{3}{10} =$

24 $\dfrac{2}{7} \times \dfrac{7}{8} \times \dfrac{1}{8} =$

Standards-Based Math • 7–8 © 2004 Creative Teaching Press

Multiplying Mixed Numbers

FACTORS AND FRACTIONS

Change to improper fractions. Multiply numerators and denominators. Simplify.

$$1\frac{2}{9} \times 3\frac{3}{4} = \frac{11}{9} \times \frac{15}{4} = \frac{11 \times 15}{9 \times 4} = \frac{165}{36} = 4\frac{21}{36} = 4\frac{7}{12}$$

Solve.

1 $4\frac{1}{3} \times 5\frac{3}{8}$

2 $2\frac{2}{3} \times 4\frac{3}{4}$

3 $3\frac{1}{3} \times 4\frac{5}{8}$

4 $5\frac{3}{8} \times 4\frac{1}{4}$

5 $3\frac{7}{8} \times 1\frac{7}{9}$

6 $4\frac{2}{3} \times 5\frac{5}{8}$

7 $9\frac{1}{3} \times 3\frac{4}{5}$

8 $4\frac{4}{7} \times 1\frac{15}{16}$

9 $3\frac{1}{5} \times 2\frac{1}{4}$

10 $4\frac{1}{3} \times 1\frac{1}{2}$

11 $\frac{3}{7} \times 5\frac{1}{4}$

12 $\frac{1}{2} \times 2\frac{1}{4}$

13 $\frac{3}{4} \times 24$

14 $17 \times \frac{1}{4}$

15 $6\frac{2}{3} \times 2\frac{1}{4}$

16 $2\frac{4}{7} \times \frac{5}{9}$

17 $7 \times 2\frac{1}{7}$

18 $2 \times 6\frac{2}{5}$

19 $9\frac{3}{4} \times 2\frac{2}{13}$

20 $\frac{5}{8} \times 1\frac{4}{5} \times 1\frac{7}{9}$

21 $\frac{1}{3} \times \frac{1}{7} \times 4\frac{2}{3}$

Dividing Fractions

FACTORS AND FRACTIONS

Multiply the dividend by the reciprocal of the divisor. Simplify.

$$\frac{3}{4} \div \frac{1}{2} = \frac{3 \times 2}{4 \times 1} = \frac{6}{4} = 1\frac{1}{2}$$

Solve.

1 $\dfrac{7}{8} \div \dfrac{2}{6} =$

2 $\dfrac{1}{9} \div \dfrac{2}{4} =$

3 $\dfrac{4}{7} \div \dfrac{4}{5} =$

4 $\dfrac{1}{3} \div \dfrac{2}{5} =$

5 $\dfrac{3}{5} \div \dfrac{7}{9} =$

6 $\dfrac{3}{8} \div \dfrac{4}{7} =$

7 $\dfrac{6}{9} \div \dfrac{2}{5} =$

8 $\dfrac{4}{8} \div \dfrac{5}{7} =$

9 $\dfrac{2}{4} \div \dfrac{5}{6} =$

10 $\dfrac{4}{7} \div \dfrac{3}{4} =$

11 $\dfrac{4}{6} \div \dfrac{2}{5} =$

12 $\dfrac{2}{8} \div \dfrac{1}{3} =$

13 $\dfrac{3}{7} \div \dfrac{3}{9} =$

14 $\dfrac{2}{8} \div \dfrac{6}{7} =$

15 $\dfrac{5}{6} \div \dfrac{6}{5} =$

16 $6 \div \dfrac{1}{2} =$

17 $7 \div \dfrac{1}{3} =$

18 $\dfrac{3}{5} \div \dfrac{1}{4} =$

19 $\dfrac{1}{3} \div \dfrac{1}{9} =$

20 $\dfrac{3}{8} \div \dfrac{3}{4} =$

21 $12 \div \dfrac{3}{4} =$

Standards-Based Math • 7–8 © 2004 Creative Teaching Press

Dividing Mixed Numbers

FACTORS AND FRACTIONS

$$\frac{1}{3} \div 5 = \frac{1}{3} \div \frac{5}{1} = \frac{1}{3} \times \frac{1}{5} = \frac{1}{15}$$

Solve.

1 $8\frac{1}{6} \div 2\frac{1}{3} =$

2 $11\frac{1}{9} \div \frac{5}{6} =$

3 $3\frac{1}{21} \div 2\frac{2}{7} =$

4 $4\frac{4}{5} \div 1\frac{4}{5} =$

5 $\frac{3}{10} \div 2\frac{2}{5} =$

6 $9\frac{3}{4} \div 2\frac{1}{6} =$

7 $1\frac{3}{4} \div 4\frac{2}{3} =$

8 $12\frac{3}{5} \div \frac{9}{10} =$

9 $16\frac{2}{3} \div 6\frac{1}{4} =$

10 $15\frac{3}{4} \div 3\frac{1}{2} =$

11 $7\frac{1}{3} \div \frac{5}{6} =$

12 $10\frac{2}{5} \div 2\frac{3}{5} =$

13 $5\frac{1}{2} \div \frac{1}{6} =$

14 $7\frac{5}{6} \div \frac{5}{6} =$

15 $9\frac{3}{4} \div 1\frac{5}{8} =$

16 $15 \div 2\frac{1}{12} =$

17 $1\frac{3}{5} \div 8 =$

18 $\frac{7}{8} \div 1\frac{1}{6} =$

19 $5\frac{1}{4} \div \frac{1}{2} =$

20 $21 \div 2\frac{1}{3} =$

21 $3\frac{4}{7} \div \frac{5}{7} =$

Name _____ Date _____

Multistep Problems

FACTORS AND FRACTIONS

Solve.

1 Henry has a doggy daycare business. He uses an average of 22 lb. of dog food every day. He buys 575 lb. of kibble. For about how many days will this last? Will there be any left over? If so, about how much?

2 If Henry buys kibble from the grocery store, the price is $0.75/lb. If he buys the kibble directly from the manufacturer, he pays $0.35/lb. How much did he save when he bought 575 lb.?

3 Henry charges $5/hr if he cares for a dog less than 30 hours a week. If he cares for a dog more than thirty hours a week, he charges a flat rate of $125.00 for the whole week. Dog owners may not leave their dogs with him for more than 50 hours a week. If you want Henry to watch your dog for 28 hours a week, would you save more money paying by the hour or paying Henry's full-time rate?

4 Last month Henry had 5 full-time dogs. He had hourly dogs for the following times each week: 12 hours, 8 hours, 22 hours, and 15 hours. There were four weeks in the month. Using the rate information from problem 3, calculate his gross earnings (total before costs).

5 Last month Henry used 450 lb. of dog food at $0.35/lb. He spent $25 on parking at the dog park. He spent $40 on a waste disposal service. He spent $75 on a professional cleaning service for the kennels. He sent 7% of his *gross* earnings to the government for taxes. Using his gross earnings from problem 4, calculate his profit (amount earned after costs are subtracted from total earnings).

Standards-Based Math • 7–8 © 2004 Creative Teaching Press

Name _____   Date _____

Along the Way

FACTORS AND FRACTIONS

Complete.

Across

2. ___ numbers are the whole numbers greater than zero.

5. A ____ is a whole number that is multiplied by another whole number to equal a product.

7. The ___ tells the total number of parts into which the whole is divided.

8. The ___ ___ ___ is the multiple with the least value that is held in common by two or more numbers.

9. ___ and fractions can represent some portion of a whole.

10. ___ numbers go on and on past the point where they can be calculated.

11. The ___ tells the portion of the whole that is referenced by the fraction.

Down

1. The ___ ___ ___ is the factor with the greatest value that is held in common by two or more numbers.

3. ___ numbers can be expressed as a terminating or repeating decimal.

4. The ___ number has exactly two factors: the number itself and 1.

6. A ___ number is a natural number that has three or more factors.

Standards-Based Math • 7–8 © 2004 Creative Teaching Press

Name _____ Date _____

Altogether

FACTORS AND FRACTIONS

Change to percent.

1 $\dfrac{2}{6}$
      2 $\dfrac{7}{27}$
      3 $\dfrac{14}{20}$
      4 $\dfrac{6}{14}$

Multiply.

5 $\dfrac{3}{4} \times \dfrac{7}{8} =$
   6 $\dfrac{5}{7} \times \dfrac{1}{5} =$
   7 $\dfrac{3}{5} \times \dfrac{1}{5} =$

8 $\dfrac{2}{7} \times \dfrac{7}{8} =$
   9 $\dfrac{4}{5} \times \dfrac{2}{7} \times \dfrac{1}{3} =$
   10 $\dfrac{2}{7} \times \dfrac{7}{8} \times \dfrac{1}{8} =$

11 $3\dfrac{1}{5} \times 2\dfrac{1}{4} =$
   12 $2 \times 6\dfrac{2}{5} =$
   13 $\dfrac{1}{3} \times \dfrac{1}{7} \times 4\dfrac{2}{3} =$

Divide.

14 $\dfrac{1}{9} \div \dfrac{2}{4} =$
   15 $\dfrac{4}{7} \div \dfrac{4}{5} =$
   16 $\dfrac{3}{8} \div \dfrac{4}{7} =$

17 $\dfrac{5}{6} \div \dfrac{6}{5} =$
   18 $7 \div \dfrac{1}{3} =$
   19 $4\dfrac{4}{5} \div 1\dfrac{4}{5} =$

20 $5\dfrac{1}{2} \div \dfrac{1}{6} =$
   21 $\dfrac{7}{8} \div 1\dfrac{1}{6} =$
   22 $12\dfrac{3}{5} \div \dfrac{9}{10} =$

Standards-Based Math • 7–8 © 2004 Creative Teaching Press

Variables

ALGEBRA

A variable is a letter or symbol that stands for a number.

$7 + n$ ◄——————— variable

Variables can be helpful when you want to solve a word problem.

Charles bought 5 new cards for a total of twenty-five cards in his collection. How many cards did he have before the purchase? You could draw a model:

— + ☐☐☐☐☐ = ☐☐☐☐☐
☐☐☐☐☐
☐☐☐☐☐
☐☐☐☐☐
☐☐☐☐☐

But it is faster to write an equation with a variable.

Original cards ————————➤ $n + 5 = 25$

Draw a model. Then write an equation. Use variable n.

1 Henry has three books in the Up and 'Attem Boys series. He buys a few more to give him a total of eight books in the series. How many books did he buy?

2 Kylie drove 200 miles today for a total of 1,000 miles traveled on this trip. How many miles had she drive on the trip before today?

Write an equation for the problem. Use variable x.

3 Max and Ben invite some friends over to play. Now there are 7 children playing at the house. How many friends came over?

4 Cora pulled another batch of cookies out of the oven. Placing the 12 cookies on the rack to cool, she said, "There! That's 48 cookies altogether!" How many cookies were done before this batch?

5 2 children are playing on the monkey bars. Another group arrives to make 8 children. How many children arrived in the second group?

6 A mini-mac truck carries 1,000 lb. of gravel. The mighty-mac truck can carry $\frac{1}{3}$ as much. How many pounds of gravel can the mighty-mac truck carry?

Name _____ Date _____

Associative and Commutative Properties

ALGEBRA

> The **Commutative Property** says you can switch the order of two numbers and still get the same answer.
> 2 + 8 = 8 + 2
> (5)(3) = (3)(5)

Solve the problem in the table. Then write two equations for the problem on a separate piece of paper.

$\times$	2	4	$\frac{1}{2}$	-3	5
$\frac{1}{3}$					
6					
$\frac{3}{5}$					
5					
8					

> The **associative property** says the way three numbers are grouped for only addition or multiplication will not effect the outcome.
> 2 + (4 + 3)   2(4•6)
> (2 + 4) + 3   (2•4)6

Rewrite the problem to demonstrate the associative property. Solve.

1 $(2 \times 3) \times 8$

2 $12 + (4 + 3)$

3 $(5 \cdot 4)3$

4 $8 + (12 + 3)$

5 $(20 + 4) + 18$

6 $(4 \times 3) \times 5$

7 $2 \times (3 \times 7)$

8 $25 + (48 + 36)$

9 $(21 + 14) + 9$

10 $4 \times (3 \times 5)$

11 $203 + (404 + 362)$

12 $3 \times (4 \times 4)$

13 $(298 + 45) + 36$

14 $(2 \times 9) \times 6$

Standards-Based Math • 7–8 © 2004 Creative Teaching Press

Distributive Property

Algebra

The **distributive property** says that when a number is multiplied by the sum or difference of two other numbers, the first number can be distributed to both of those two numbers and multiplied by each of them separately.

$$a(b + c) = ab + ac$$
$$5(4 + 3) = 5 \times 4 + 5 \times 3$$
$$5(7) = 20 + 15$$
$$35 = 35$$

Rewrite the equation to distribute the multiplication. Solve both sides to check your work.

1 $3(5 + 2) =$

2 $^{-}4(5 + 3) =$

3 $2(5 + 6) =$

4 $(4 - 2)5 =$

5 $7(6 + 2) =$

6 $5(18 - 9) =$

7 $(2 + {}^{-}6)8 =$

8 $(5 + 6)3 =$

9 $7(7 - 3) =$

10 $9(20 + 3)$

11 $5(8 + 2)$

12 $6(8 + 4)$

13 $3(22 + 3)$

14 $6(7 + 9)$

15 $^{-}5(8 + 4)$

16 $8(12 + ({}^{-}2))$

17 $9(7 + 8)$

18 $6(30 + 33)$

19 $4(8 + 8)$

20 $9(58 + ({}^{-}4))$

21 $8(14 + ({}^{-}7))$

22 $6(3 + (6))$

23 $12(6 + ({}^{-}20) + 2)$

24 $9(2 + 3)$

25 $6(6 + 2)$

26 $9(4 + 8)$

27 $5({}^{-}7 + 9)$

Order of Operations

Algebra

Parenthesis
Exponents
Multiplication
Division
Addition
Subtraction

Solve.

1 $7 + (8 - 2) \times 2$

2 $\dfrac{4^2}{2^2} =$

3 $(5 + 3)^2$

4 $(4 \times 2) - 3$

5 $\dfrac{(8 + 2)^2}{(4 + 1)^2} + 4 =$

6 $(50 + 6) \div 7 =$

7 $2(3 + 5)^2 - 3$

8 $31 + (3)(10)$

9 $[2(2 + 4) + 4] \bullet 3 + 2$

10 $5 + \dfrac{9 + 5}{3 + 4}$

11 $\dfrac{1}{2}(4 + 4)$

12 $(3 + 4)(7 + 1)$

13 $\left(\dfrac{2^3}{16}\right)(6)$

14 $\dfrac{1}{3}(6 + 4 + 2) + 8$

Name _____ _____ Date _____

Inverse Operations

ALGEBRA

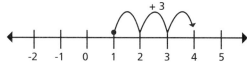
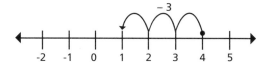
Match each operation and value to its inverse.

1 + 5

2 – 10

3 × 3

4 ÷ 6

5 + (⁻4)

6 ÷ 0.5

7 – 1.07

8 + 143

9 + $\frac{6}{2}$

10 ÷ 8

11 × (⁻4)

12 + 19

A. ÷ (⁻4)

B. + 1.07

C. + 4

D. – 143

E. ⁻5

F. × 6

G. + 10

H. – 19

I. × 8

J. ÷ 3

K. × 0.5

L. – 3

Inverse Operations: Addition

ALGEBRA

To solve an equation with one variable, use inverse operations until all known values are on the same side of the = and the unknown value is on the other side of the =. Whatever you do to one side, you must do to the other.

$$x + 6 = 3x - 2$$
$$ - 6 - 6$$
$$x = 3x - 8$$
$$x - 3x = 3x - 3x - 8$$
$$\frac{-2x}{-2} = \frac{-8}{-2}$$
$$x = 4$$

Solve.

1 $x + 4 = 12$

2 $a + 6 = 24 + 9$

3 $y + 5 = 18$

4 $z + (5 + 3) = 50$

5 $x + (4 - 2) = 12$

6 $a + (3 \times 5) = 100$

7 $15 + x = 100$

8 $9 + a = 27$

9 $(7 + 9) + y = 75$

10 $(6 \times 4) + b = 60$

11 $(9 - 4) + x = 25$

12 $8 + c = 8$

13 $x + 2.7 = 7.7$

14 $x + 12 = 52$

15 $29 + n = 32$

16 $23 + r = 18$

17 $34 = n + 13$

18 $t + 19 = 39$

19 $52 + x = 100$

20 $x + \dfrac{1}{4} = 1$

Standards-Based Math • 7–8 © 2004 Creative Teaching Press

Inverse Operations: Subtraction

A L G E B R A

$$32 = x - 2$$
$$32 + 2 = x - 2 + 2$$
$$34 = x$$

Solve.

1 $x - 4 = 12$

2 $a - 6 = 15 + 9$

3 $y - 5 = 20$

4 $z - (7 + 3) = 85$

5 $x - (9 - 2) = 78$

6 $a - (9 \times 5) = 100$

7 $^-15 + x = 100$

8 $^-6 + a = 27$

9 $y - (9 \div 3) = 19$

10 $b - 0.5 = 5.9$

11 $x - \dfrac{1}{3} = 3\dfrac{2}{3}$

12 $^-4 + c = 0$

13 $a - 65 = 7$

14 $x - 58 = 16$

15 $x - 4\dfrac{3}{8} = 6$

16 $c - 42 = 67$

17 $y - 4.8 = 9.2$

18 $b - 4\dfrac{1}{4} = 12\dfrac{2}{3}$

19 $c - 6.05 = 848.95$

20 $b - 55 = 83$

Inverse Operations: Multiplication

ALGEBRA

$$30 = 4x - 2$$
$$32 = 4x$$
$$\frac{32}{4} = \frac{4x}{4}$$
$$8 = x$$

Solve.

1 $6x = 30$

2 $9x = 90$

3 $45 = 5x$

4 $65 = 3x - 5$

5 $27 = 6x - 3$

6 $7x = 42$

7 $40 = 5x + 5$

8 $28 = 4x$

9 $8x = 72$

10 $3x = 45$

11 $100x = 10,000$

12 $x^2 = 25$

13 $6x = 42$

14 $4x = 28$

15 $7x = 49$

16 $5x = 280$

17 $5.4x = 21.6$

18 $4.5x = 9$

19 $4.8x = 36$

20 $13x = 169$

Standards-Based Math • 7–8 © 2004 Creative Teaching Press

Inverse Operations: Division

ALGEBRA

$$32 = \frac{x}{2}$$

$$32 \times 2 = \frac{x}{2} \times 2$$

$$64 = x$$

Solve.

1 $\dfrac{x}{7} = 6$ **2** $x \div 12 = 4$ **3** $x \div 8 = 6$ **4** $x \div 5 = 9$ **5** $\dfrac{x}{4} = 5$

6 $x \div 10 = 12$ **7** $\dfrac{x}{3} = 9$ **8** $\dfrac{x}{5} + 6 = 21$ **9** $3x \div 2 = 9$ **10** $\dfrac{x}{8} + 9 = 20$

11 $2x \div 5 = 10$ **12** $\dfrac{x}{5} - 8 = 3$ **13** $\dfrac{x}{4} = 9$ **14** $\dfrac{x}{5} = 7$ **15** $x \div 3 = 10$

16 $x \div 5 = 24$ **17** $\dfrac{x}{9} = 15$ **18** $\dfrac{x}{3} = 8$ **19** $\dfrac{x}{5} = 33$ **20** $x \div 25 = 4$

Variables 2

Algebra

Solve for x.

1 $3(x + 4) = 27$

2 $(6x)(5) = 60$

3 $X + 3 = 15$

4 $4x + 15 = 39$

5 $\dfrac{4x}{4} = 7$

6 $6(3x) = 18$

7 $10^x = 1,000$

8 $^-8(^-5x) = 800$

9 $\dfrac{(3)(8)}{x} = 6$

10 $\dfrac{1}{2}\, x = 50$

11 $x^2 = 81$

12 $\dfrac{x}{7} - 4 = 3$

13 $^-2100x = 6300$

14 $11x = 33$

15 $x - 6 = ^-18$

16 $x - (^-6) = ^-7$

17 $x \div 3 = ^-3$

18 $x + 5 = ^-2$

19 $6x = 18$

20 $4x = 36$

Standards-Based Math • 7–8 © 2004 Creative Teaching Press

Inequalities

ALGEBRA

$$7x > 49$$
$$\frac{7x}{7} > \frac{49}{7}$$
$$x > 7$$

$$X - 3 < 9$$
$$X - 3 + 3 < 9 + 3$$
$$X < 12$$

Solve for X.

1 $3x - 7 < 2$

2 $-\frac{1}{3} x < 48$

3 $x + \frac{5}{6} < -\frac{1}{2}$

4 $x + 37 > 98$

5 $x - 53 < 141$

6 $\frac{x}{3} > \frac{1}{3}$

7 $\frac{11}{2} x < 3\frac{2}{3}$

8 $\frac{x}{-5} > 11$

9 $-\frac{7}{8} x > -\frac{7}{8}$

10 $\frac{x}{4} < -45$

11 $x + \frac{5}{8} > \frac{3}{4}$

12 $(18 - 12)x > 96$

13 $x + 8 > 6$

14 $x + \frac{1}{4} \geq 1\frac{1}{4}$

15 $-20 + x \leq 24$

16 $3\frac{1}{3} \leq \frac{2}{9} + x$

17 $x - 3 \leq 1$

18 $-2.4 < x - 0.6$

19 $-4 \leq 1 + x$

20 $-3 > x + 1$

Graphing Inequalities

ALGEBRA

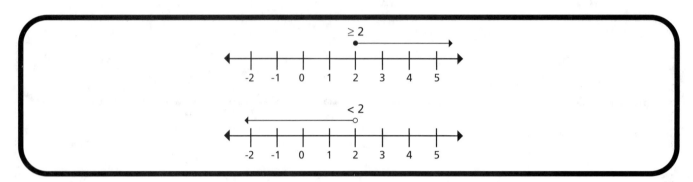

Represent the following values.

1 >4

2 >2.5

3 <3$\frac{3}{4}$

4 <2 and ≥5

Solve. Graph the solution.

5 y + 7 < 11

6 n + 2 ≥ 4

7 z + 3 < 8

8 3x − 7 < 2

9 $\frac{x}{3}$ ≤ −15

10 (15 − 12)x ≥ 96

Standards-Based Math • 7–8 © 2004 Creative Teaching Press

Name _____ Date _____

Lines and Line Segments

Geometry

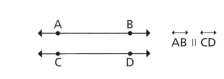 $\overleftrightarrow{AB} \parallel \overleftrightarrow{CD}$

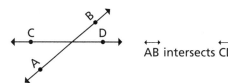 $\overleftrightarrow{AB}$ intersects $\overleftrightarrow{CD}$

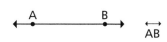 $\overleftrightarrow{AB}$

$\overleftrightarrow{AB} \perp \overleftrightarrow{CD}$

$\overline{AB}$

Use the label that best describes the illustration.

1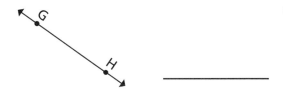

_____

2

_____

3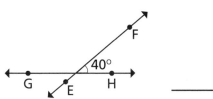

_____

4

_____

5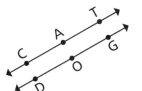

_____

6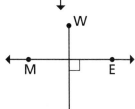

_____

7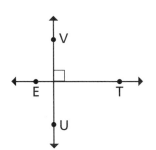

_____

8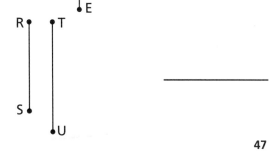

_____

Standards-Based Math • 7–8 © 2004 Creative Teaching Press

Angles

GEOMETRY

Match each illustration to its definition.

①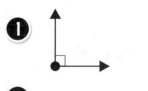

②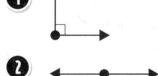

③

④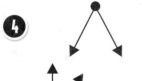

⑤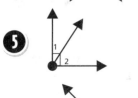

⑥

A. acute angle: measures less than 90°

B. right angle: measures exactly 90°

C. supplementary angle: two angles that create a 180° angle

D. obtuse angle: measures more than 90°

E. complementary angle: two angles that create a 90° angle

F. straight angle: measures exactly 180°

Draw hands on the clock that illustrate the kind of angle indicated.

⑦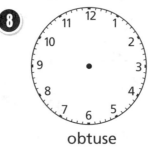

straight

⑧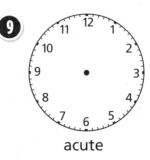

obtuse

⑨

acute

Use a protractor to measure each angle.

⑩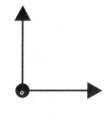

_____

⑪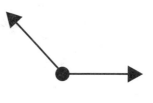

_____

⑫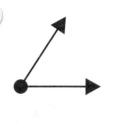

_____

Angle Measurement

GEOMETRY

The sum of the interior angles of a triangle is 180°.

Find the measure of angle A.

1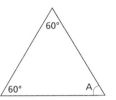

2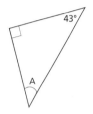

3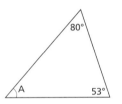

4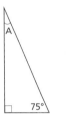

5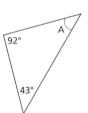

6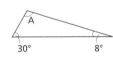

7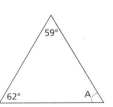

8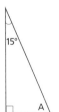

9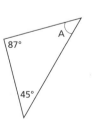

10

11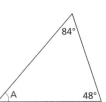

12

Name _____ Date _____

Geometry Terms

GEOMETRY

Complete.

Across

2. A __ is a flat surface that extends endlessly in all directions.
3. A __ is a geometric object that has no dimensions, only a location.
4. Lines that never intersect and are in the same plane are called __.
5. Two faces of a solid meet at the __.
6. A line segment that extends from one edge of a circle to the other and passes through its center is also its ____.
8. A line segment that extends from one edge of the circle to the other edge of the circle is a ____.
9. A __ is a three-dimensional geometric figure.
11. ___ lines meet at one point.
14. A __ is a flat surface of a geometric solid.

Down

1. The __ is a point equal distance from the end points of a line.
3. Lines that intersect at a right angle are __ lines.
7. A line segment that extends from the center of a circle to its edge is also its ___.
10. Two rays joined at a common endpoint form an ___.
12. A line that intersects with a circle at only one point of its edge is a ____.
13. Part of a line with only one endpoint is called a ___.

Standards-Based Math • 7–8 © 2004 Creative Teaching Press

Name _____ _____ Date _____

Polygons

GEOMETRY

A polygon's prefix names its number of sides. Most polygon prefixes are Greek in origin.

Number	=	Prefix	Polygon
treis, tria	3	tri-	triangle
quattor*	4	quadri-, quart-	quadrilateral
pente	5	penta-	pentagon
hexa	6	hex-	hexagon
hepta	7	hept-	heptagon
okto	8	oct-	octagon
ennea	9	ennea	nonagon
deka	10	dec-	decagon
dodeka	12	dodec-	dodecagon

*Latin in origin, as is *lateral,* which means sides.

Label each polygon.

_____

_____

_____

_____

_____

_____

_____

_____

_____

_____

Name _____ Date _____

Similar and Congruent

GEOMETRY

> **Similar shapes** are the same shape, but not the same size.
> **Congruent shapes** are the same shape and size.

Label each set of figures **similar** or **congruent**.

1

_____

2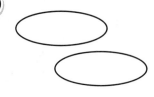

_____

3

_____

4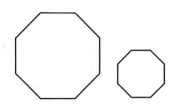

_____

5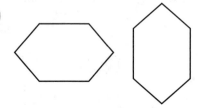

_____

6

_____

7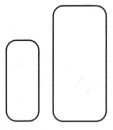

_____

8

_____

9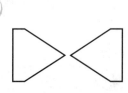

_____

10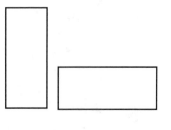

_____

11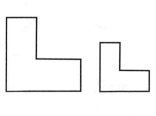

_____

12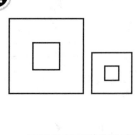

_____

Standards-Based Math • 7–8 © 2004 Creative Teaching Press

Similar Triangles

GEOMETRY

Similar triangles have congruent angles. The perimeters and each corresponding side share the same ratio. This is called the ratio of scale.

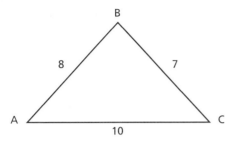
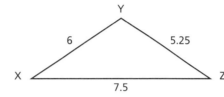

$$\frac{\overline{AB}}{\overline{XY}} = \frac{8}{6} = 1\frac{1}{3} = 1.\overline{33}$$

$$\frac{\overline{BC}}{\overline{YZ}} = \frac{7}{5.25} = 1.\overline{33}$$

$$\frac{\overline{AC}}{\overline{XZ}} = \frac{10}{7.5} = 1.\overline{33}$$

$$\Delta ABC \sim \Delta XYZ \sim \Delta HIJ$$

Use a known ratio to find the length of unknown sides.

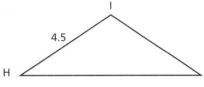

$$\frac{\overline{YZ}}{\overline{IJ}} = \frac{5.25}{1.33} = 3.9375 \approx 3.9$$

$$\frac{\overline{XZ}}{\overline{HJ}} = \frac{7.5}{1.33} = 5.625 \approx 5.6$$

$$\overline{IJ} = 3.9$$

$$\overline{HJ} = 5.6$$

Evaluate the similar triangles. Find the ratio of scale. Use this information to fill in all the missing side measurements. Use what you know about triangles to fill in the missing angle measurements.

1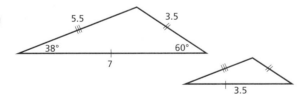

2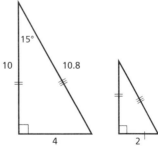

3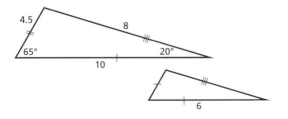

4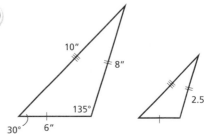

Name _____ Date _____

Similar Figures

Geometry

Similar figures also have equal angles and sides that share the same scale ratio.

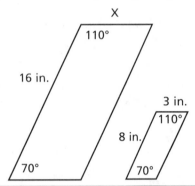

X
110°
16 in.
70°

3 in.
110°
8 in.
70°

$$\frac{16}{x} = \frac{8}{3}$$

$$8x = 3 \times 16$$

$$\frac{8x}{8} = \frac{48}{8}$$

$$x = 6 \text{ in.}$$

Calculate the ratio and complete the missing side measurements.

1
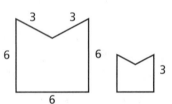
3   3
6   6
6
3

2
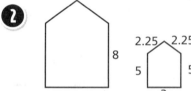
8
2.25   2.25
5   5
3

3
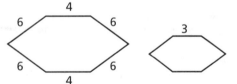
4
6   6
6   6
4
3

4
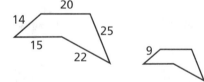
20
14
25
15
22
9

5
5
5.5
3
6   4

6
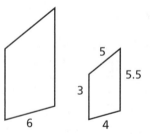
6
5
2

7
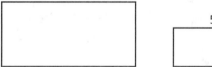
10
5   10

8
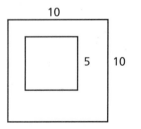
15
4.5   5
5

Standards-Based Math • 7–8 © 2004 Creative Teaching Press

Parts of a Circle

GEOMETRY

Match.

1 radius

2 chord

3 tangent

4 diameter

5 arc

6 semi-circle

A. A line segment with both endpoints on the circle and that passes through the center.

B. An arc that is one-half the circumference of a circle.

C. The distance from the center to a point on the circle.

D. A line perpendicular to the radius that touches only one point on the circle.

E. A line segment that connects two points on a curve.

F. A portion of the circumference of a circle.

Use colored pencils.

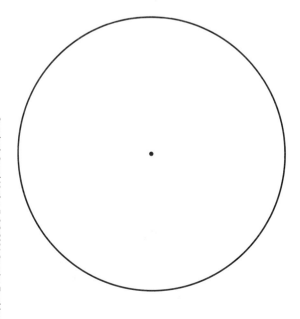

Use a red pencil to draw radius $\overline{AB}$.
Use a green pencil to draw chord $\overline{CD}$.
Use a purple pencil to show the diameter $\overline{EAF}$.
Use a yellow pencil to trace the arc $\overset{\frown}{CD}$.
Use an orange pencil to trace the semi-circle $\overset{\frown}{EBF}$.
Use a blue pencil to show tangent $\overleftrightarrow{GH}$.

Standards-Based Math • 7–8 © 2004 Creative Teaching Press

Name _____ Date _____

Geometric Solids

GEOMETRY

Cylinders are solids with two circular bases and a curved surface.

Cones are solids with a circular base, a curved surface, and one vertex.

Spheres have a curved surface where every point is an equal distance from the center.

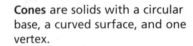

Prisms are solids with two parrallel, congruent polygon bases and rectangular faces that connect them.

Pyramids are solids with a polygon base, and triangular faces that come to a point.

Name the geometric solid. Label the type of polygon of the base(s).

1

_____

2

_____

3

_____

4

_____

5

_____

6

_____

7

_____

8

_____

9

_____

10

_____

11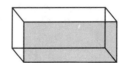

_____

12 How is a pyramid like a cone?

_____

_____

Standards-Based Math • 7–8 © 2004 Creative Teaching Press

Name _____ Date _____

Spatial Relationships

GEOMETRY

Cones are solids with a circular base, a curved surface, and one vertex.

Match the name of the solid to the figure that could be assembled to form it.

1 dodecahedron
_____

2 pentagonal pyramid
_____

3 cylinder
_____

4 triangular prism
_____

5 tetrahedron
_____

6 rectangular prism
_____

7 triangular prism
_____

8 cube
_____

9 rectangular pyramid
_____

10 cone
_____

11 octahedron
_____

A

B

C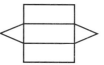

D

E

F

G

H

I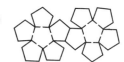

J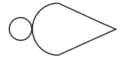

K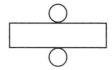

Name _____ Date _____

Angles, Circles, and Solids

Geometry

Name the type of geometric figure.

1
A   B

_____

2

_____

3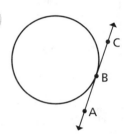

_____

4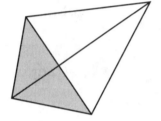

_____

5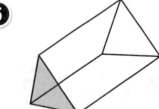

_____

6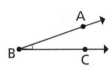

_____

7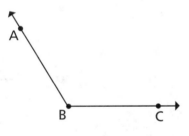

_____

8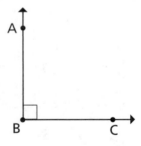

_____

9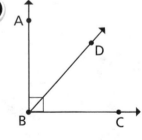

_____

10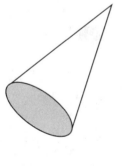

_____

11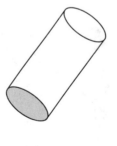

_____

12

_____

Standards-Based Math • 7–8 © 2004 Creative Teaching Press

The X, Y Axis

GEOMETRY

The **Cartesian** plane is made up of a vertical line called the **Y-axis** and a horizontal line called the **X-axis**. They are perpendicular to each other and meet at point zero.

Any point on the plane can be described with a pair of numbers called an **ordered pair**. The first number always refers to where the point lines up with the X-axis and the second number always refers to where the point lines up with the Y-axis.

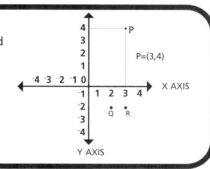

Write the ordered pair for each point.

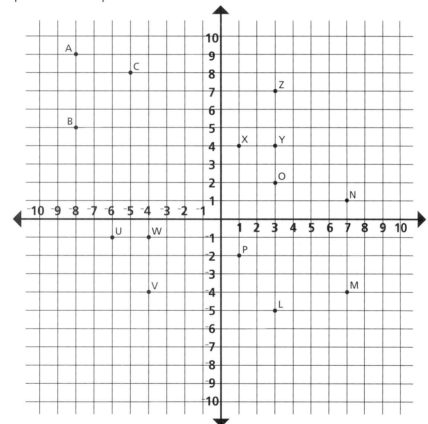

1 A =

2 B =

3 C =

4 L =

5 M =

6 N =

7 O =

8 P =

9 U =

10 V =

11 W =

12 X =

13 Y=

14 Z =

Using a Cartesian Plane

GEOMETRY

Write the ordered pair at which you stop.

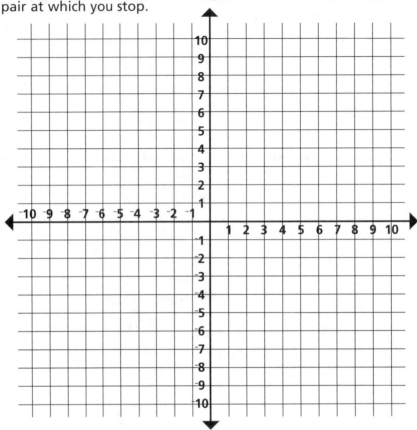

1 Start at point (2, 3). Move 2 points to the right, 2 to the left, and 3 to the left. Final point: _____

2 Start at point (5, 4). Move 4 points to the right, 2 to the left, and 8 to the left. Final point: _____

3 Start at point (6, ⁻2). Move 8 points to the left, and 5 up. Final point: _____

4 Start at point (2, 8). Move 2 points up, 1 down, and three to the right. Final point: _____

5 Start at point (⁻3, ⁻5). Move 0 points left, 4 down, and 5 left. Final point: _____

6 Start at point (6, 3). Move 0 points down, 2 up, and 5 down. Final point: _____

7 Start at point (⁻5, 0). Move 2 points right, 9 up, and 5 down. Final point: _____

8 Start at point (⁻3, 4). Move 8 points to the left, 5 right, and 7 up. Final point: _____

Standards-Based Math • 7–8 © 2004 Creative Teaching Press

Reflections Across the X-Axis

Geometry

When a figure is reflected across the X-axis, the x coordinate does not change. The y coordinate switches from + to – or – to +.

(2, 3) (2, ‾3)
(3, 5) (3, ‾5)
(4, 4) (4, ‾4)

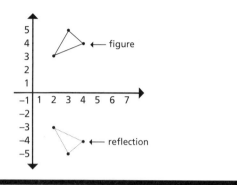

Record the reflected coordinates on the table. Then, draw an XY axis on graph paper and check your work by plotting both sets of points.

①

Figure	Reflection
(2, ‾8)	
(2, ‾2)	
(4, ‾2)	
(4, ‾6)	
(7, ‾6)	
(7, ‾8)	

②

Figure	Reflection
(1, 3)	
(1, 8)	
(5, 8)	
(5, 6)	
(4, 6)	
(4, 5)	
(5, 5)	
(5, 3)	

③

Figure	Reflection
(‾3, 3)	
(‾5, 6)	
(‾3, 9)	
(‾5, 9)	
(‾6, 7)	
(‾7, 9)	
(‾9, 9)	
(‾7, 6)	
(‾9, 3)	
(‾7, 3)	
(‾6, 5)	
(‾5, 3)	

④

Figure	Reflection
(‾6, ‾2)	
(‾4, ‾4)	
(‾6, ‾5)	
(‾4, ‾8)	
(‾8, ‾8)	
(‾10, ‾5)	
(‾8, ‾4)	
(‾10, ‾2)	

Name _____ Date _____

Reflections Across the Y-Axis

GEOMETRY

When a figure is reflected across the Y-axis, the y coordinate does not change. The x coordinate switches from + to – or – to +.

(2, 3)   (⁻2, 3)
(3, 5)   (⁻3, 5)
(4, 4)   (⁻4, 4)

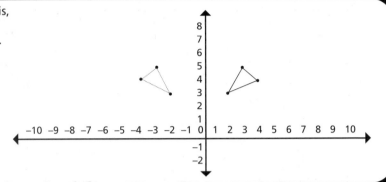

Record the reflected coordinates on the table. Then, draw an XY axis on graph paper and check your work by plotting both sets of points.

1

Figure	Reflection
(⁻2, 6)	
(⁻2, 8)	
(⁻5, 8)	
(⁻5, 7)	
(⁻7, 7)	
(⁻7, 4)	
(⁻5, 4)	
(⁻5, 6)	

2

Figure	Reflection
(⁻3, 3)	
(⁻3, 6)	
(⁻4, 5)	
(⁻4, 8)	
(⁻7, 7)	
(⁻7, 5)	
(⁻6, 6)	
(⁻6, 3)	
(⁻7, 2)	
(⁻4, 2)	

3

Figure	Reflection
(2, 4)	
(2, 6)	
(4, 9)	
(5, 7)	
(8, 9)	
(8, 6)	
(5, 6)	
(6, 4)	

4

Figure	Reflection
(4, 5)	
(6, 7)	
(9, 7)	
(11, 5)	
(8, 5)	
(10, ⁻2)	
(5, ⁻2)	
(7, 5)	

Standards-Based Math • 7–8 © 2004 Creative Teaching Press

Translations Along the X-Axis

GEOMETRY

Adding or subtracting a constant value to the X-axis of each point of a figure will translate it along the X-axis.

x, y	(x – 6), y
(3, 4)	(⁻3, 4)
(3, 7)	(⁻3, 7)
(5, 6)	(⁻1, 6)
(7, 7)	(1, 7)
(7, 4)	(1, 4)
(5, 5)	(⁻1, 5)

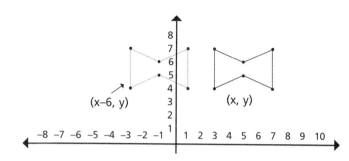

Record the translated coordinates on the table. Then, draw an XY axis on graph paper and check your work by plotting both sets of points.

①

x + 7	
Figure	**Translation**
(⁻9, 4)	
(⁻6, 8)	
(⁻5, 4)	

②

x – 5	
Figure	**Translation**
(5, 5)	
(6, 10)	
(8, 8)	
(8, 5)	

③

x – 4	
Figure	**Translation**
(3, ⁻9)	
(7, ⁻3)	
(8, ⁻6)	
(6, ⁻9)	

④

x + 8	
Figure	**Translation**
(⁻6, 2)	
(⁻7, 4)	
(⁻6, 6)	
(⁻7, 8)	
(⁻6, 10)	
(⁻4, 10)	
(⁻5, 8)	
(⁻4, 6)	
(⁻5, 4)	
(⁻4, 2)	

Name _____ Date _____

Translations Along the Y-Axis

GEOMETRY

Adding or subtracting a constant value to the Y-axis of each point of a figure will translate it along the Y-axis.

x, y	x, y−7)
(7, 3)	(7, ⁻4)
(5, 5)	(5, ⁻2)
(5, 9)	(5, 2)
(8, 7)	(8, 0)
(8, 5)	(8, ⁻2)

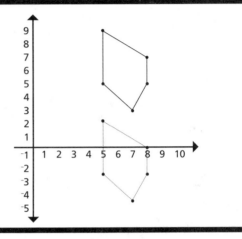

Record the translated coordinates on the table. Then, draw an XY axis on graph paper and check your work by plotting both sets of points.

1

y − 5	
Figure	**Translation**
(2, 5)	
(2, 7)	
(8, 5)	
(7, 4)	
(4, 4)	

2

y + 6	
Figure	**Translation**
(⁻10, ⁻3)	
(⁻9, ⁻7)	
(⁻7, ⁻6)	
(⁻5, ⁻4)	
(⁻8, ⁻4)	

3

y − 10	
Figure	**Translation**
(3, 5)	
(3, 7)	
(5, 8)	
(7, 7)	
(9, 8)	
(9, 5)	

4

y + 8	
Figure	**Translation**
(⁻10, ⁻2)	
(⁻7, ⁻2)	
(⁻5, ⁻4)	
(⁻3, ⁻2)	
(⁻3, ⁻7)	
(⁻5, ⁻5)	
(⁻7, ⁻7)	
(⁻10, ⁻7)	

Standards-Based Math • 7-8 © 2004 Creative Teaching Press

Name _____ Date _____

Oblique Translations

GEOMETRY

Adding or subtracting a constant value to the X-axis and Y-axis of each point of a figure will translate it to another location on the grid.

x, y	(x – 4), (y – 7)
(7, 3)	(3, ⁻4)
(5, 5)	(1, ⁻2)
(5, 9)	(1, 2)
(8, 7)	(4, 0)
(8, 5)	(4, ⁻2)

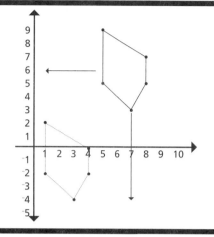

Record the translated coordinates on the table. Then, draw an XY axis on graph paper and check your work by plotting both sets of points.

①

(x + 5), (y – 5)	
Figure	**Translation**
(2, 4)	
(2, 5)	
(6, 5)	
(6, 4)	

②

(x – 7), (y + 6)	
Figure	**Translation**
(1, 3)	
(1, 7)	
(3, 5)	
(3, 2)	

③

(x – 3), (y – 10)	
Figure	**Translation**
(12, 5)	
(9, 3)	
(12, 2)	
(15, 3)	

④

(x + 8), (y + 8)	
Figure	**Translation**
(⁻4, ⁻2)	
(⁻2, ⁻4)	
(⁻3, ⁻5)	
(⁻2, ⁻7)	
(⁻4, ⁻7)	
(⁻5, ⁻5)	
(⁻4, ⁻4)	
(⁻6, ⁻2)	

Name _____ Date _____

Transformations

GEOMETRY

Look at the diagram. Name the kind of transformation performed.

①

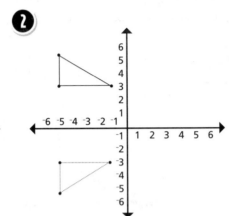

_____

_____

②

_____

_____

③

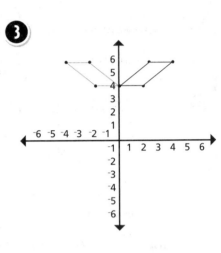

_____

_____

④

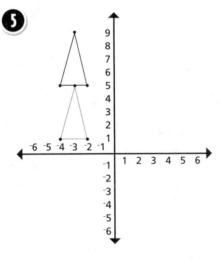

_____

_____

⑤

_____

_____

⑥

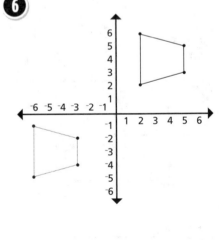

_____

_____

Standards-Based Math • 7–8 © 2004 Creative Teaching Press

Perimeter

GEOMETRY

Perimeter is a measurement of the distance around an object.

Find the perimeter.

1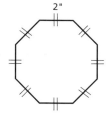
2"

_____

2 5 ft.
6 ft.

_____

3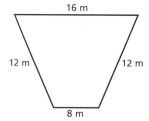
16 m
12 m   12 m
8 m

_____

4
9 cm

_____

5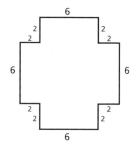
6
2   2
2   2
6   6
2   2
2   2
6

_____

6
8 km

_____

7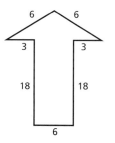
6   6
3   3
18   18
6

_____

8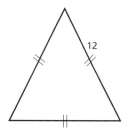
12

_____

9
8

_____

10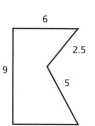
6
2.5
9
5

_____

11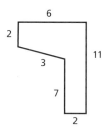
6
2
3
11
7
2

_____

12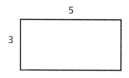
5
3

_____

Name _____ Date _____

Area

GEOMETRY

Area is a measurement of the total surface expressed in square units.

Parallelogram: A = b × h

Rectangle: A = l × w

Square: A = s²

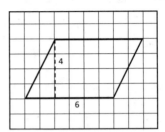

6 × 4 = 24 sq. units

Calculate the area.

1  2.5

2 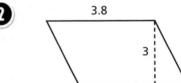 3.8 3

3 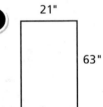 21" 63"

4   2 cm

5 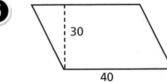 30 40

6 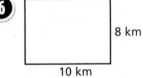 8 km 10 km

7 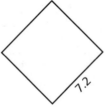 7.2

8 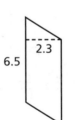 2.3 6.5

9 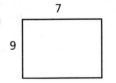 7 9

10 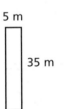 5 m 35 m

11   40

12 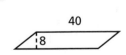 40 8

Standards-Based Math • 7–8 © 2004 Creative Teaching Press

Name _____ Date _____

Area of a Triangle and Trapezoid

GEOMETRY

Triangle: A = $\frac{1}{2}$ bh

A = $\frac{1}{2}$ (6)(8) = 24 sq. units

Trapezoid: A = $\frac{1}{2}$ h(b^1 + b^2)

A = $\frac{1}{2}$ 12(5 + 15) = 120 sq. units

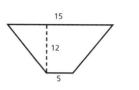

Add or subtract. Then use the inverse operation to check your work.

1
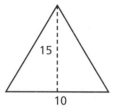
15
10

2

50
20

3
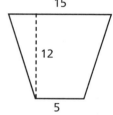
15
12
5

4
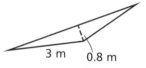
3 m   0.8 m

5
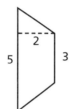
2
5   3

6
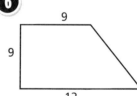
9
9
12

7
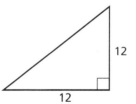
12
12

8
1.5 in.
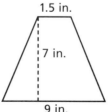
7 in.
9 in.

9

70
220

10
5
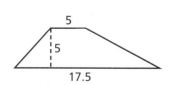
5
17.5

11
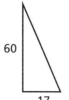
60
17

12

14 mi.
16 mi.

Circumference

GEOMETRY

The **circumference** of a circle is the distance around the outside of the circle. Circumference = the perimeter of a circle.

$C = \pi d$ d = diameter
$\pi = 3.14$ r = radius
$C = \pi d$ $d = 2r$

$C = \pi(2)(9) = \pi18 = (3.14)(18) = 56.52$ in.

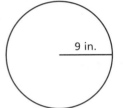
9 in.

Find the circumference.

1 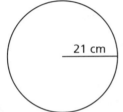 21 cm

2 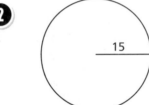 15

3 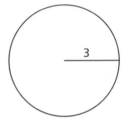 3

4 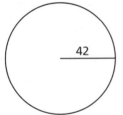 42

5 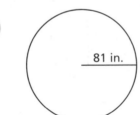 81 in.

6 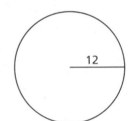 12

7 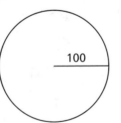 100

8 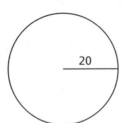 20

9 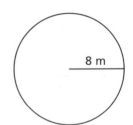 8 m

10 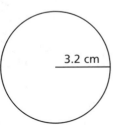 3.2 cm

11 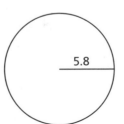 5.8

12 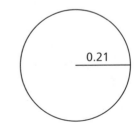 0.21

Name _____   Date _____

Area of Circles

GEOMETRY

A = πr²

r = radius

π = 3.14

A = π9² = π81 = 254.34 sq. m

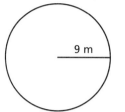
9 m

Find the area of each circle.

1
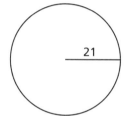
21

2
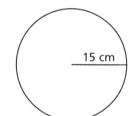
15 cm

3
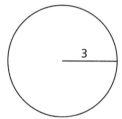
3

4
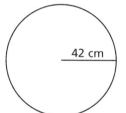
42 cm

5
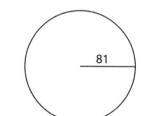
81

6
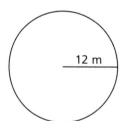
12 m

7
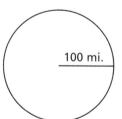
100 mi.

8
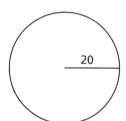
20

9
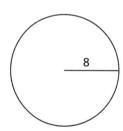
8

10
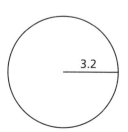
3.2

11
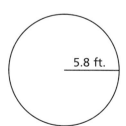
5.8 ft.

12
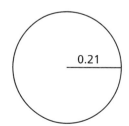
0.21

Area and Circumference of Circles

GEOMETRY

$$A = \pi r^2 \qquad C = \pi d$$

Find the area and the circumference for each circle.

1 A _____

C _____

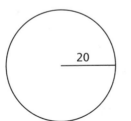

2 A _____

C _____

3 A _____

C _____

4 A _____

C _____

5 A _____

C _____

6 A _____

C _____

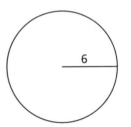

7 A _____

C _____

8 A _____

C _____

9 A _____

C _____

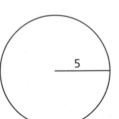
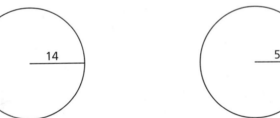
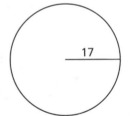
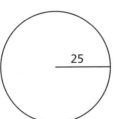
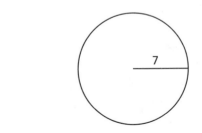

Standards-Based Math • 7–8 © 2004 Creative Teaching Press

Points, Shapes, and Solids

GEOMETRY

Each pair of figures is similar. Calculate for x.

1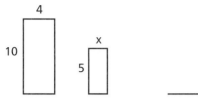

_____

2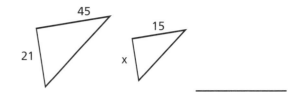

_____

3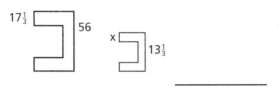

_____

4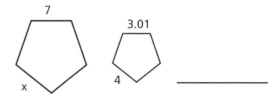

_____

Write the coordinates for point P.

5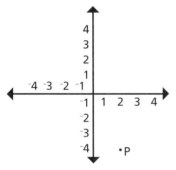

6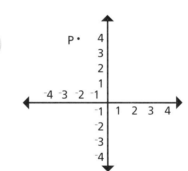

7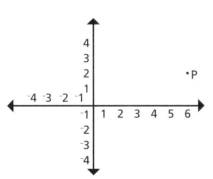

8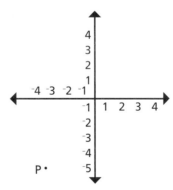

Find the area and perimeter.

9

A = _____
P = _____

10

A = _____
P = _____

Name _____ Date _____

Ratios

RATIO AND PROPORTION

Ratio: Represents the relationship of two quantities. Ratios can be written:

$$\frac{1}{3}, \text{ 1 to 3, or 1:3}$$

box height = 20 inches   box width = 10 inches

$$\frac{20}{10} = \frac{2}{1}, \text{ 2 to 1 or 2:1}$$

Tell in your own words what each ratio describes. The first one is done for you.

1 $\dfrac{80 \text{ pieces of candy}}{20 \text{ students}}$ =

2 $\dfrac{50" \text{ of rain}}{5 \text{ months}}$ =

3 $\dfrac{350 \text{ words}}{5 \text{ minutes}}$ =

4 pieces of candy per students

_____

_____

4 $\dfrac{320 \text{ miles}}{14.5 \text{ gal. of gas}}$ =

5 $\dfrac{48 \text{ seeds}}{6 \text{ pots}}$ =

6 $\dfrac{100 \text{ m}}{14.5 \text{ sec.}}$ =

_____

_____

_____

7 $\dfrac{500 \text{ stars}}{10 \text{ flags}}$ =

8 $\dfrac{300 \text{ min. of knitting}}{5 \text{ days}}$ =

9 $\dfrac{60 \text{ pencils}}{12 \text{ students}}$ =

_____

_____

_____

Rewrite as a ratio.

10 The school has 5 teachers for every 100 students.

11 1 out of every 3 baskets was scored by Jimmy.

12 2 inches of every foot of wood is wasted by the builder.

Standards-Based Math • 7–8 © 2004 Creative Teaching Press

Rates

RATIO AND PROPORTION

> **Rate:** Represents the relationship of two quantities that have different units of measure. A rate is a kind of ratio.
>
> Trip distance = 250 miles   Trip time = 5 hours
>
> $$\frac{250}{5} = \frac{50}{1} \text{ or 50 miles per hour}$$

Write the rate in fraction form.

1 1,200 m in 3 days

2 $260 in 20 days

3 300 flowers planted in 20 days

4 480 beads for 12 necklaces

5 $810 in 9 hours

6 2,500 fish in 50 tanks

7 232 mi. in 16 hrs.

8 258 km in 3 hours

9 15 yards of fabric for 5 dresses

10 $50 for 2 classes

11 8 mi. in 2 days

12 160 eggs for 40 omelets

Name _____ Date _____

Proportions

RATIO AND PROPORTION

Proportion: An equation that sets two ratios equal to each other.

$$\frac{a}{4} = \frac{1}{2}$$

$$\frac{2}{4} = \frac{1}{2}$$

Solve.

1 $\dfrac{a}{5} = \dfrac{8}{10}$

2 $\dfrac{1}{3} = \dfrac{4}{d}$

3 $\dfrac{2}{10} = \dfrac{c}{5}$

4 $\dfrac{24}{80} = \dfrac{x}{60}$

5 $\dfrac{1}{b} = \dfrac{4}{16}$

6 $\dfrac{50}{100} = \dfrac{10}{d}$

7 $\dfrac{3}{b} = \dfrac{6}{16}$

8 $\dfrac{a}{9} = \dfrac{2}{6}$

9 $\dfrac{a}{5} = \dfrac{8}{10}$

10 $\dfrac{6}{24} = \dfrac{2}{d}$

11 $\dfrac{20}{x} = \dfrac{16}{36}$

12 $\dfrac{10}{16} = \dfrac{x}{56}$

13 $\dfrac{33}{99} = \dfrac{21}{x}$

14 $\dfrac{48}{80} = \dfrac{x}{5}$

15 $\dfrac{5}{n} = \dfrac{25}{45}$

Cross Multiplication
RATIO AND PROPORTION

Cross product property: Given two equivalent ratios, $\frac{a}{b}$ and $\frac{c}{d}$, the cross product property states that $a \times d = b \times c$.

$$\frac{a}{2} = \frac{3}{4}$$
$$a \times 4 = 3 \times 2$$
$$a \times 4 = 6$$
$$a \times 4 \div 4 = 6 \div 4$$
$$a = \frac{3}{2}$$

Use the cross product property to solve the equations.

1 $\dfrac{4}{6} = \dfrac{3}{d}$

2 $\dfrac{6}{b} = \dfrac{3}{5}$

3 $\dfrac{1}{b} = \dfrac{4}{16}$

4 $\dfrac{1}{3} = \dfrac{c}{24}$

5 $\dfrac{a}{7} = \dfrac{14}{49}$

6 $\dfrac{a}{3} = \dfrac{2}{6}$

7 $\dfrac{4}{10} = \dfrac{6}{d}$

8 $\dfrac{11}{12} = \dfrac{22}{d}$

9 $\dfrac{15}{4} = \dfrac{x}{20}$

10 $\dfrac{6}{a} = \dfrac{4}{32}$

11 $\dfrac{x}{6} = \dfrac{55}{66}$

12 $\dfrac{5}{a} = \dfrac{45}{54}$

13 $\dfrac{x}{4} = \dfrac{18.4}{27.6}$

14 $\dfrac{3x}{10} = \dfrac{9}{10}$

15 $\dfrac{4}{12} = \dfrac{3}{x}$

Proportions 2

RATIO AND PROPORTION

Solve each proportion for x.

1 $\dfrac{2}{18} = \dfrac{x}{36}$

2 $\dfrac{7}{x} = \dfrac{14}{16}$

3 $\dfrac{3}{5} = \dfrac{x}{20}$

4 $\dfrac{5}{x} = \dfrac{25}{45}$

5 $\dfrac{7}{9} = \dfrac{63}{x}$

6 $\dfrac{1}{4} = \dfrac{5}{x}$

7 $\dfrac{x}{9} = \dfrac{4}{18}$

8 $\dfrac{7}{8} = \dfrac{21}{x}$

9 $\dfrac{x}{3} = \dfrac{15}{18}$

10 $\dfrac{x}{9} = \dfrac{64}{72}$

11 $\dfrac{3}{x} = \dfrac{18}{30}$

12 $\dfrac{15}{12} = \dfrac{x}{4}$

13 $\dfrac{4}{5} = \dfrac{x}{20}$

14 $\dfrac{x}{45} = \dfrac{10}{15}$

15 $\dfrac{8}{36} = \dfrac{2}{x}$

16 $\dfrac{3}{x} = \dfrac{6}{15}$

17 $\dfrac{7}{6} = \dfrac{x}{18}$

18 $\dfrac{5}{25} = \dfrac{1}{x}$

19 $\dfrac{x}{40} = \dfrac{3}{8}$

20 $\dfrac{10}{x} = \dfrac{5}{6}$

Standards-Based Math • 7–8 © 2004 Creative Teaching Press

Name _____ Date _____

Unit Price

Ratio and Proportion

Which is the better deal?

$$\frac{2}{\$0.59} = \frac{1}{x} \qquad x = 29.5¢$$

$$\frac{5}{1.29} = \frac{1}{x} \qquad x = 26¢$$

2 for 59¢

5 for $1.29

Find the unit price.

1 3 for 63¢

2 5 for $15.00

3 8 for $34.00

4 7 for $9.03

5 4 for $94.40

6 50 for $18.95

7 64 for $18.50

8 5 for $14.95

9 12 for $21.00

Find the unit price. Then tell which choice is the better deal.

10 5 lb. of apples for $5.95 or 8 lb. of avocados for $28.00?

11 3 pairs of socks for $4.99 or 12 pairs of socks for $14.99?

12 5 lb. of honey for $8.95 or 10 lb. of honey for $18.00?

13 $17.22 for 8 gallons of gas or $38.85 for 18 gallons?

14 $2.95 for 16 oz. of milk or $3.99 for 1 gallon?

15 $38.95 for 8 oz. of caviar or $25.95 for 4 oz.?

Ratio and Proportion

RATIO AND PROPORTION

Write each ratio as a fraction in simplest terms.

1 196 to 7

2 65 out of 105

3 112 : 140

4 0.11 to 1.21

5 19 : 76

6 18 to 27

Solve each proportion for x.

7 $\dfrac{x}{9} = \dfrac{4}{18}$

8 $\dfrac{7}{8} = \dfrac{21}{x}$

9 $\dfrac{x}{3} = \dfrac{15}{18}$

10 $\dfrac{x}{9} = \dfrac{64}{72}$

11 $\dfrac{3}{x} = \dfrac{18}{30}$

12 $\dfrac{15}{12} = \dfrac{x}{4}$

13 $\dfrac{x}{40} = \dfrac{3}{8}$

14 $\dfrac{10}{x} = \dfrac{5}{6}$

Solve each word problem.

15 The Smith family travels 1,000 miles in four days. If they drive about the same number of miles each day, about how many miles did they travel in one day?

16 If two lb. of ground beef cost $3.90, how much should five lb. cost?

Standards-Based Math • 7–8 © 2004 Creative Teaching Press

Name _____ Date _____

Percent

DATA ANALYSIS AND PROBLEM SOLVING

$$\frac{1}{2} = 0.5$$
$$0.5 \times 100 = 50\%$$
$$\frac{1}{2} = 0.5 = 50\%$$

Use mental math to solve.

1 $\frac{1}{5} =$

2 $\frac{3}{12} =$

3 $\frac{15}{50} =$

4 $\frac{24}{60} =$

5 $\frac{1}{20} =$

6 $\frac{2}{40} =$

7 $\frac{4}{5} =$

8 $\frac{24}{30} =$

9 $\frac{1}{10} =$

10 $\frac{4}{4} =$

11 $\frac{5}{8} =$

12 $\frac{2}{5} =$

13 $\frac{3}{20} =$

14 $\frac{6}{75} =$

15 $\frac{8}{128} =$

16 $\frac{11}{165} =$

17 $\frac{12}{108} =$

18 $\frac{22}{25} =$

19 $\frac{63}{100} =$

20 $\frac{14}{50} =$

Standards-Based Math • 7–8 © 2004 Creative Teaching Press

Fractions, Decimals, and Percents

DATA ANALYSIS AND PROBLEM SOLVING

$$\frac{1}{6} = 0.16\overline{6}$$

$$0.166 \approx 0.17$$
$$0.17 \times 100 = 17\%$$

$$\text{So, } \frac{1}{6} = 0.166 \approx 17\%$$

Covert the fraction to a decimal, round the decimal to the nearest hundredths, then change it to a percent.

1 $\dfrac{2}{6} =$ _____ $\times 100 =$ _____

2 $\dfrac{5}{18} =$ _____ $\times 100 =$ _____

3 $\dfrac{4}{12} =$ _____ $\times 100 =$ _____

4 $\dfrac{5}{9} =$ _____ $\times 100 =$ _____

5 $\dfrac{4}{21} =$ _____ $\times 100 =$ _____

6 $\dfrac{5}{24} =$ _____ $\times 100 =$ _____

7 $\dfrac{7}{27} =$ _____ $\times 100 =$ _____

8 $\dfrac{4}{33} =$ _____ $\times 100 =$ _____

9 $\dfrac{8}{24} =$ _____ $\times 100 =$ _____

10 $\dfrac{5}{54} =$ _____ $\times 100 =$ _____

11 $\dfrac{20}{60} =$ _____ $\times 100 =$ _____

12 $2\dfrac{7}{25} =$ _____ $\times 100 =$ _____

13 $\dfrac{7}{10} =$ _____ $\times 100 =$ _____

14 $\dfrac{14}{70} =$ _____ $\times 100 =$ _____

15 $\dfrac{100}{1,000} =$ _____ $\times 100 =$ _____

16 $\dfrac{3}{5} =$ _____ $\times 100 =$ _____

17 $\dfrac{33}{50} =$ _____ $\times 100 =$ _____

18 $\dfrac{66}{75} =$ _____ $\times 100 =$ _____

Standards-Based Math • 7–8 © 2004 Creative Teaching Press

More Fractions, Decimals, and Percents

DATA ANALYSIS AND PROBLEM SOLVING

$$34\% = 0.34 = \frac{34}{100} = \frac{17}{50}$$

Complete the chart.

	Percent	Decimal	Fraction		Percent	Decimal	Fraction
1.	45%			**15.**	50%		
2.	210%			**16.**		0.44	
3.	0.4%			**17.**			4/9
4.		0.123		**18.**		0.75	
5.			$\frac{3}{10}$	**19.**	9%		
6.			$\frac{1}{3}$	**20.**	80%		
7.		0.125		**21.**			3/200
8.	1%			**22.**		0.78	
9.			$\frac{1}{11}$	**23.**			7/25
10.	14%			**24.**		0.95	
11.			$\frac{7}{8}$	**25.**	32%		
12.			$\frac{2}{3}$	**26.**		0.18	
13.		$0.1\overline{6}$		**27.**	98%		
14.		0.25		**28.**			1/5

Name _____ Date _____

Percent of a Number

Data Analysis and Problem Solving

$$
\begin{array}{r}
4\% \text{ of } 20 = \\
0.04 \times 20 = \\
2\ 0 \\
\times\ 0.\ 0\ 4 \\
\hline
0.\ 8\ 0 \\
4\% \text{ of } 20 = 0.8
\end{array}
$$

Round to the nearest percent.

1 33% of $12.00 = _____

2 98% of 250 = _____

3 120% of 625 = _____

4 9% of 60 = _____

5 20% of 80 = _____

6 2% of 1200 mL = _____

7 82% of 6 m = _____

8 25% of $550.00 = _____

9 6% of 380 = _____

10 32% of 297 = _____

11 25% of $60.00 = _____

12 90% of 120 = _____

13 50% of 200 = _____

14 25% of 80 = _____

15 33% of $24.00 = _____

16 55% of 28 = _____

Standards-Based Math • 7–8 © 2004 Creative Teaching Press

Name _____ Date _____

Data Analysis

DATA ANALYSIS AND PROBLEM SOLVING

It's time for the Corn on the Cob Day Parade! Kylie takes information from each group in the parade. She makes the following chart.

Group	No. of People Riding on Float	No. of Animals Participating	Approx. Weight of Float	Throwing Candy?
4-H	8	24	1.5 tons	No
Larson's Tire	4	0	1.75 tons	Yes
ABC Daycare	12	2	1.5 tons	Yes
Central High Marching Band	6 (56 marching)	0	2.25 tons	No
Bob's Feed and Grain	8	2	3 tons	Yes
Girl Scouts	24	4	1.6 tons	Yes
Boy Scouts	18	0	1.5 tons	Yes
Central Farm Co-op	6	2	2.25 tons	Yes
Acme Hardware	4	1	2 tons	No

Solve.

1 Kylie wants to arrange the floats so that the smallest floats are at the front and the largest are at the back. Assume that the weight of the float indicates its relative size. How would you order the floats?

2 The judges ask Kylie to tell them the average number of animals participating. She thinks they really want to know the most common number of animals on the floats. Calculate the average, then calculate the mode. Then, write a sentence telling why Kylie thinks the mode will be more useful than the mean.

3 The parade committee has to pay a special tax to the city if the average weight of the floats is over 2 tons. Will they have to pay this tax this year?

4 What is the range of people riding on the floats? (Do not include the 56 members of the marching band, only the 6 members of Homecoming Court riding with them.)

Mean, Median, Mode, and Range

DATA ANALYSIS AND PROBLEM SOLVING

Mean: the average
Median: the middle number of the data when the data is arranged in order
Mode: the data value that appears most frequently
Range: the difference between the greatest and least values of the data set

Find the mean, median, mode, and range for each set of data.

1 6, 9, 8, 9, 12, 15, 12, 6, 9, 7, 15, 12

mean :_____   median :_____   mode :_____   range :_____

2 0.43, 0.76, 0.46, 0.55, 0.25, 0.32, 0.37, 0.50

mean :_____   median :_____   mode :_____   range :_____

3 89, 93, 51, 64, 91, 103, 46, 64, 82, 123, 112, 99

mean :_____   median :_____   mode :_____   range :_____

4 16, 23, 18, 30, 19, 37, 23, 30, 33, 16, 30

mean :_____   median :_____   mode :_____   range :_____

5 17,423; 13,678; 19,555; 18,000; 16,894

mean :_____   median :_____   mode :_____   range :_____

6 $24.00, $160.00, $78.00, $44.00, $48.00, $36.00, $44.00

mean :_____   median :_____   mode :_____   range :_____

7 88%, 73%, 94%, 84%, 73%, 62%

mean :_____   median :_____   mode :_____   range :_____

8 77, 111, 57, 98, 107, 44, 35, 107, 66

mean :_____   median :_____   mode :_____   range :_____

Standards-Based Math • 7–8 © 2004 Creative Teaching Press

Reading Tables

DATA ANALYSIS AND PROBLEM SOLVING

Babies Born May 1st at Animosa Hospital

Name	M/F	Eye Color	Length	Weight	Time
Phan, M.	M	Brown	21 in.	8 lb. 2 oz.	1:45 am
Scotts, A.	M	Brown	20 in.	7 lb. 12 oz.	3:15 am
Prescott, B.	F	Brown	19 in.	6 lb. 14 oz.	9:19 am
Mendez, R.	M	Brown	20.5 in.	7 lb. 13 oz.	9:25 am
Peterson, A.	F	Blue	18 in.	5 lb. 2 oz.	2:00 pm
Peterson, B.	F	Blue	18.25 in.	5 lb. 6 oz.	2:01 pm

Use the chart to answer the questions.

1 How many children were born at Animosa Hospital on May 1st?

_____

2 What was the most common eye color?

_____

3 Based on the results, which two babies might be twins?

_____

4 The data is organized by

_____. In order from _____.

5 What was the average weight of these babies?

_____

6 What was the average length?

_____

7 What is the weight difference between the heaviest and lightest infants?

_____

8 Are there more boys or girls?

_____

Name _____ Date _____

Reading Line Graphs

DATA ANALYSIS AND PROBLEM SOLVING

Line Graphs are useful for showing how something changes over time. Like a scatter plot, they have an X-axis and a Y-axis. The X-axis typically shows time, and the Y-axis typically shows the numbers that correlate to the data.

Read the graphs and answer the questions.

①

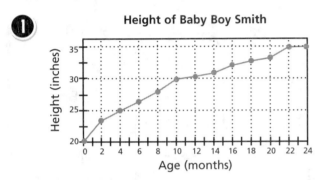

Height of Baby Boy Smith

What was Baby Boy Smith's height when he was 10 months old?
How much had he grown since his previous measurement?
How much did he grow altogether over the two years?

②

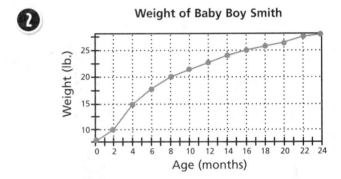

Weight of Baby Boy Smith

During which interval was there the greatest amount of weight gain?
During which interval was there the least change in Baby Boy Smith's weight?

③

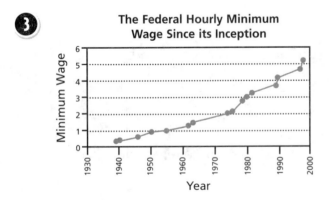

The Federal Hourly Minimum Wage Since its Inception

What does this graph describe?
Was there ever a period when the minimum wage decreased?
What key piece of information is missing from one of the labels?

④

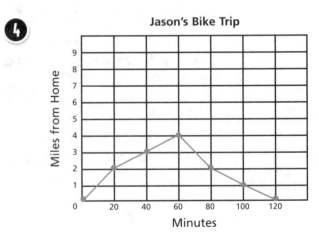

Jason's Bike Trip

How long did Jason's bike trip last?
There is a long upgrade, or hill, during part of Jason's ride. Judging from the data, during which period was he riding up it and during which period was he riding down it?

Standards-Based Math • 7–8 © 2004 Creative Teaching Press

Creating a Circle Graph

DATA ANALYSIS AND PROBLEM SOLVING

Find the degrees of your pie wedge by multiplying the percentage of the circle times 360°.

McIntosh	70	8%

.08 × 360° ≈ 29°

Complete the chart. Then, use the data to create a pie graph.

Apples Sold

Apple Type	Number Sold	Percent of Whole
Red Delicious	200	
Golden Delicious	350	
Granny Smith	175	
Jonathan	50	
Winesap	55	
McIntosh	70	

Hint: Calculate the total number of apples sold first.

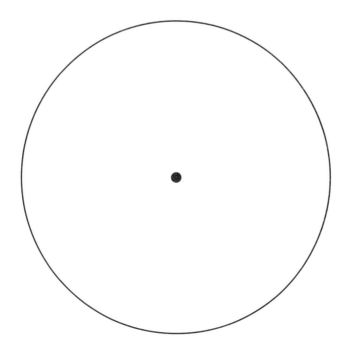

Name _____ Date _____

Scatter Plots

DATA ANALYSIS AND PROBLEM SOLVING

A scatter plot is only for pairs of related data. Assign one data set the X-values and one data set the Y-values. Then, plot as you normally would on an XY axis.

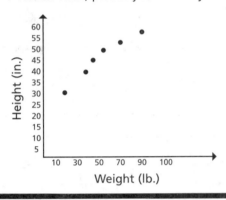

Height (in.)	Weight (lb.)
32	20
38	40
45	45
48	55
53	70
56	90

Construct scatter plots for each of the charts below on graph paper. Complete them with the information provided and give each scatter plot a title.

① Puppy Growth

Height	Weight
≤6	1.5 lb.
≤8	3 lb.
≤9.5	8 lb.
≤12	11 lb.
≤13	15 lb.
≤15	20 lb.
≤18	25 lb.

② Weed Growth

Number of Weeds in 8 ft. Square	Inches of Rain Previous Week
6	0.02
7	0.09
9	0.11
14	0.52
20	0.66
45	0.78
49	0.23

③ Age by Grade

Age	Grade
5	K
6	1
8	2
7	2
9	4
7	1
10	5

④ Age by Number of Siblings

Age	Number of Siblings
9	0
10	1
7	2
6	0
9	1
5	5
7	3

⑤ Which graph above shows the least amount of relationship between the two data sets?

Standards-Based Math • 7–8 © 2004 Creative Teaching Press

Name _____ Date _____

Reading Graphs

DATA ANALYSIS AND PROBLEM SOLVING

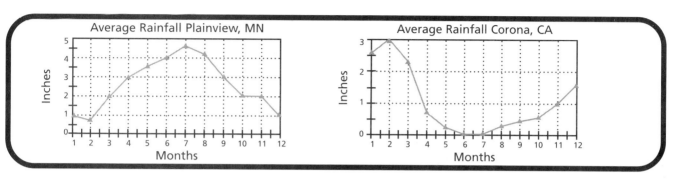

Use the graphs to answer the questions. Numbers represent each month (e.g., May is the 5th month).

1 Which city gets more rain in July?

2 Which city gets less rain overall?

3 What period of time has the greatest reduction in rainfall for each city?

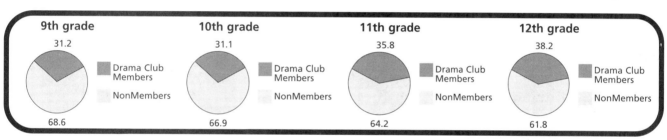

Use the graphs to answer the questions.

4 Which grade level has the greatest percentage of Drama Club Members?

5 What is the increase in the percentage of students who join Drama Club between the 9th grade and 12th grade years?

6 What is the average percentage of Drama Club members per grade level, according to the graphs?

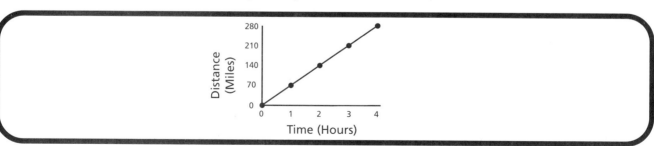

Use the graph to answer the questions.

7 How far is the object traveling every hour?

8 Does its speed ever change?

9 Based on the graph, how many miles do you predict the object will have traveled after five hours?

Name _____ Date _____

The Bottom Line
DATA ANALYSIS AND PROBLEM SOLVING

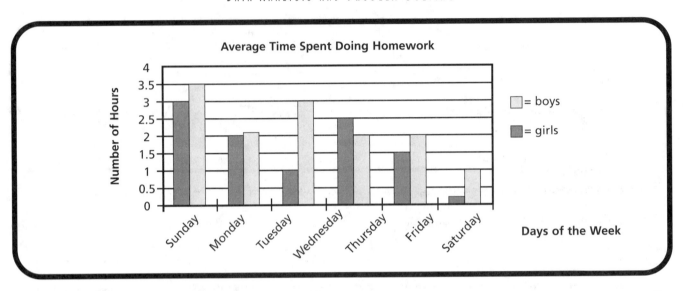

Use the graph to answer the questions.

1 Which night of the week is most likely to be spent doing homework by both boys and girls?

2 How many fewer hours are girls likely to spend on Tuesday night homework than on Sunday night homework?

3 What is the average amount of time that boys spend on homework in a week?

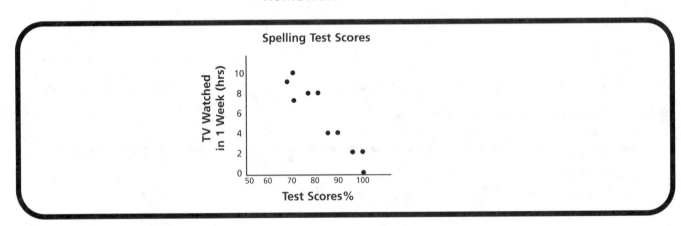

Use the graph to answer the questions.

4 Describe the general path of the scatter plot. What relationship do you see?

5 How many pieces of data are represented on the graph?

6 Is the data clustered more toward one end or the other? That is, can you say whether more of the students polled watched a lot of TV or a little TV?

Standards-Based Math • 7–8 © 2004 Creative Teaching Press

Name _____ Date _____

Defining Integers

INTEGERS

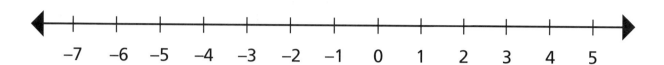

-7 -6 -5 -4 -3 -2 -1 0 1 2 3 4 5

Positive integers represent an increase of value. Integers are used most often to note a change in measurement such as an increase in weight, volume, or money.
Negative integers represent a decrease of value. A loss of weight, volume, or money might be noted with a negative integer.

Circle the greatest value. Underline the least value.

1 7, 3, -8, -9, 2

2 -5, 6, 12, -12, 4, 0

3 -4, -8, -5, -6, 0, -3

4 -2, -9, -5, -3, -13

Joe borrows $5 from his Dad for a ream of paper and $20 from his mom for a new printer ink cartridge. Altogether, what is his financial situation?

-$5.00 + -$20.00 = -$25.00

Hattie is recording the movements of a mouse across a grid. She notes that from 0 he runs 3" to the right, 8" to the left, and then returns to center.

3" + -8" + 5" = 0

Maria's cat is recovering after an illness. They are monitoring her weight very carefully. After initially losing 7 lb., she gained 2 lb. last week, and 2 lb. this week. How far is she from her starting weight?

-7 lb. + 2 lb. + 2 lb. = -4 lb.

Write an equation and solve.

5 Cara monitors a snail in her aquarium. She notes that this morning it crawled 1" up the glass. A few hours later it crawled another 2" up the glass. Later, it crawled 4" down the glass. How far is the snail from where it started?

6 Henry starts the day with $30. He and his sister go to the fair. He pays for his own admission of $10, but fails to notice that $15 falls out of his wallet while he is paying. His sister lends him $10 so that he can still enjoy the fair. Write the equation that illustrates his day's financial events.

7 Karen opens a credit account to purchase a new bed. She makes a down payment of $200 on a $1,000 bed. What is her financial situation?

8 Felicia owes her Dad $5.00. She earns $12.00 doing chores. How much did Felicia receive after her Dad took out what she owed him?

Standards-Based Math • 7–8 © 2004 Creative Teaching Press

Absolute Value

Integers

> **Absolute value:** The distance to a number from zero. It is written $|x|$ and is read "the absolute value of x."
> $$|-4| = 4 \text{ and } |4| = 4$$

Simplify

1. $|-4| =$

2. $|5| =$

3. $|8 - 4| =$

4. $|4 - 8| =$

5. $-|7| =$

6. $-|-7| =$

7. $-|-3| + {}^-3 =$

8. $|{}^-8| - |{}^-4| =$

9. $|{}^-6| \cdot |{}^-4| =$

10. $|{}^-2| \cdot |{}^-8| =$

11. $|{}^-4 \cdot 5| =$

12. $|{}^-4| \div |{}^-2| =$

13. $-\left(-|{}^-10|\right) =$

14. $|{}^-5| + |{}^-7 \cdot 3| =$

15. $-|{}^-4| + |{}^-5 \cdot 2| =$

16. $-|{}^-8| + {}^-\left(|{}^-9 \cdot 4|\right) =$

17. $-|{}^-9| - \left(|{}^-10 \cdot 7|\right) =$

18. $-|{}^-4| \cdot {}^-\left(|{}^-3 \cdot 2|\right)$

19. $\dfrac{|{}^-4|}{|{}^-16|} =$

20. $\left|\dfrac{{}^-1}{{}^-2}\right| + \left|\dfrac{{}^-1}{{}^-2}\right| =$

Standards-Based Math • 7–8 © 2004 Creative Teaching Press

Adding Integers

INTEGERS

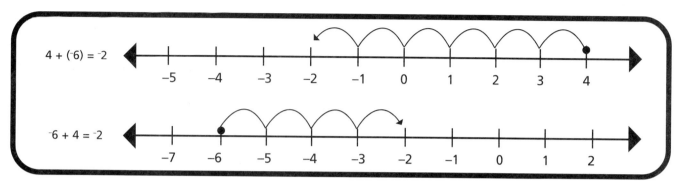

4 + (⁻6) = ⁻2

⁻6 + 4 = ⁻2

Use mental math to solve.

1 4 + (⁻8) =

2 23 + (⁻16) =

3 ⁻36 + (⁻23) =

4 21 + (⁻35) =

5 ⁻5 + 17 =

6 ⁻7 + (⁻7) =

7 23 + 4 =

8 25 + (⁻12) =

9 ⁻42 + 16 =

10 ⁻30 + (⁻18) =

11 51 + (⁻28) =

12 9 + (⁻18) =

13 ⁻26 + (⁻28) =

14 (53) + (⁻50) =

15 10 + (⁻3) + (⁻5) + 10 =

16 8 + (⁻4) + 12 + (⁻15) =

Subtracting Integers

INTEGERS

Subtract an integer by adding its opposite.

$$^-8 - 5 =$$
$$^-8 + ^-5 = ^-13$$

$$^-8 - ^-5 =$$
$$^-8 + 5 = ^-3$$

The negative of a negative is a positive. $-(-a) = a$

Rewrite each as an addition problem. Solve.

1 $^-72 - 70 = ^-72 +$ _____ = _____

2 $52 - (^-46) =$

3 $90 - 49 =$

4 $7 - (^-95) =$

5 $^-55 - (^-59) =$

6 $^-47 - 20 =$

Solve.

7 $57 + (^-70) =$

8 $^-32 - (^-88) =$

9 $^-84 + (^-61) =$

10 $^-18 - 39 =$

11 $21 + (^-24) =$

12 $79 - 55 =$

13 $97 + (^-99) =$

14 $68 - (^-73) =$

15 $^-92 + 43 =$

16 $49 + (^-19) =$

17 $^-18 - (^-9) =$

18 $^- 34 - 42$

19 $45 - (^-5) =$

20 $67 - (^-20) =$

Standards-Based Math • 7–8 © 2004 Creative Teaching Press

Name _____ Date _____

Exploring Linear Equations

Integers

Adding or subtracting a constant value to the X-axis of each point of a figure will translate it along the X-axis.

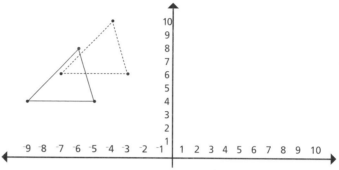

Rule = (x + 2), (y+2)	
Figure	**Translation**
(⁻9, 4)	(⁻7, 6)
(⁻6, 8)	(⁻4, 10)
(⁻5, 4)	(⁻3, 6)

Complete each table.

❶

Rule = (x + 2), (y+1)	
Figure	**Translation**
(⁻9, 4)	(⁻8, 5)
(⁻6, 8)	(⁻5, 9)
(⁻5, 4)	(⁻4, 5)

❷

Rule = _____	
Figure	**Translation**
(5, 5)	(4, 8)
(6, 10)	(5, 13)
(8, 8)	(7, 11)
(8, 5)	(7, 8)

❸

Rule = _____	
Figure	**Translation**
(3, ⁻9)	(7, ⁻11)
(7, ⁻3)	(11, ⁻5)
(8, ⁻6)	(12, ⁻8)
(6, ⁻9)	(10, ⁻11)

❹

Rule = _____	
Figure	**Translation**
(⁻6, 2)	(⁻12, 3)
(⁻7, 4)	(⁻14, 5)
(⁻6, 6)	(⁻12, 7)
(⁻7, 8)	(⁻14, 9)
(⁻6, 10)	(⁻12, 11)

❺

Rule = _____	
Figure	**Translation**
(⁻9, 4)	(⁻8, 12)
(⁻6, 8)	(⁻5, 24)
(⁻5, 4)	(⁻4, 12)

❻

Rule = _____	
Figure	**Translation**
(5, 5)	(11, 5)
(6, 10)	(13, 10)
(8, 8)	(17, 8)
(8, 5)	(17, 5)

Positive and Negative

INTEGERS

Simplify

1 $^-|-7| =$

2 $^-|^-3| + |^-3| =$

3 $|^-8| - |^-4| =$

4 $|^-6| \bullet |^-4| =$

5 $|^-2| \bullet |^-8| =$

6 $|^-4 \bullet 5| =$

7 $|^-4 \div ^-2| =$

8 $23 + 4 =$

9 $25 + (^-12) =$

10 $^-42 + 16 =$

11 $^-30 + (^-18) =$

12 $21 + (^-24) =$

13 $79 - 55 - 97 + (^-99) =$

14 $68 - (^-73) =$

Complete the table.

15

Rule = _____	
Figure	**Translation**
(⁻5, 4)	(⁻10, 5)
(⁻6, 8)	(⁻12, 9)
(⁻7, 4)	(⁻14, 5)

16

Rule = _____	
Figure	**Translation**
(3, 5)	(4, 15)
(6, 10)	(7, 30)
(7, 8)	(8, 24)
(9, 11)	(10, 33)

Standards-Based Math • 7–8 © 2004 Creative Teaching Press

Terminating Decimals

RATIONAL NUMBERS

Convert $\frac{1}{2}$ to a decimal.

Divide the numerator by the denominator. $2\overline{)1.0}^{\,0.5}$

Convert each fraction to a decimal.

1 $\frac{1}{5}$ = _____

2 $\frac{3}{12}$ = _____

3 $\frac{15}{50}$ = _____

4 $\frac{24}{60}$ = _____

5 $\frac{1}{20}$ = _____

6 $\frac{2}{40}$ = _____

7 $\frac{4}{5}$ = _____

8 $\frac{24}{30}$ = _____

9 $\frac{1}{10}$ = _____

10 $\frac{4}{4}$ = _____

11 $\frac{5}{8}$ = _____

12 $\frac{2}{5}$ = _____

13 $\frac{3}{20}$ = _____

14 $\frac{6}{75}$ = _____

15 $\frac{21}{40}$ = _____

16 $\frac{18}{90}$ = _____

17 $\frac{35}{80}$ = _____

18 $\frac{36}{80}$ = _____

19 $\frac{6}{150}$ = _____

20 $\frac{78}{800}$ = _____

Repeating Decimals

RATIONAL NUMBERS

Convert $\frac{1}{3}$ to a decimal.

Divide the numerator by the denominator.

$$\overset{\overline{0.333}}{3\overline{)1.000}}$$

$$\underline{0.9}$$
$$0.10$$
$$\underline{0.09}$$
$$0.010$$
$$0.009$$

A line over the last digit(s) of a decimal means that the decimal repeats those numbers from that point on.

Convert each fraction to a repeating decimal.

1 $\frac{2}{6} =$

2 $\frac{5}{18} =$

3 $\frac{4}{12} =$

4 $\frac{5}{9} =$

5 $\frac{4}{21} =$

6 $\frac{5}{24} =$

7 $\frac{7}{27} =$

8 $\frac{4}{33} =$

9 $\frac{8}{24} =$

10 $\frac{4}{18} =$

11 $\frac{25}{45} =$

12 $\frac{36}{54} =$

13 $\frac{36}{162} =$

14 $\frac{8}{48} =$

15 $\frac{11}{66} =$

16 $\frac{40}{72} =$

17 $\frac{80}{240} =$

18 $\frac{42}{72} =$

19 $\frac{100}{150} =$

20 $\frac{50}{540} =$

Standards-Based Math • 7–8 © 2004 Creative Teaching Press

Comparing Rational Numbers

Rational Numbers

Use <, >, or = to complete.

1 $5\dfrac{1}{2}$ ◯ $5.\overline{55}$

2 $\dfrac{22}{66}$ ◯ $\dfrac{22}{66}$

3 $^-12 \times 3$ ◯ $^-(6^2)$

4 $^-5$ ◯ $^-5\dfrac{1}{9}$

5 $^-7$ ◯ 7

6 $3\dfrac{5}{8}$ ◯ $3\dfrac{5}{12}$

7 5 ◯ 0.5

8 170 ◯ 107

9 301 ◯ 310

10 $25 + 3^2$ ◯ 34

11 1.2×1.5 ◯ 1.8

12 105 ◯ 104 ◯ 103

13 0.562 ◯ 0.0562

14 15×2 ◯ 96

15 $55 \div 11$ ◯ 4

16 $\dfrac{7}{14}$ ◯ $\dfrac{1}{2}$

17 0.2 ◯ $\dfrac{1}{5}$

18 $35 - 3^2$ ◯ $24 + 3^2$

The Density Property

RATIONAL NUMBERS

Density property: Between any two rational numbers, there is always another rational number.

Name a rational number between the two numbers of each pair.

1 4.05   4.06   **2** ⁻0.3 $\frac{1}{3}$

3 4.78   4.77   **4** 3,450   3,449

5 ⁻17.8   ⁻17.9   **6** 9   9.01

7 $\frac{5}{6}$ $\frac{1}{2}$ **8** $\frac{9}{16}$ $\frac{1}{2}$

9 ⁻8   ⁻9   **10** $\frac{4}{5}$ 1

11 890   891   **12** 45   44.9

13 3.08   3.09   **14** ⁻20   ⁻21

15 $-\frac{1}{2}$ 0.5

Scientific Notation

RATIONAL NUMBERS

Product property: When multiplying powers having the same base, add the exponents.
$$a^2 \times a^3 = a^{2+3} = a^5$$

Power property: When finding the power of a power, multiply the exponents.
$$(a^2)^3 = a^{2 \times 3} = a^6$$

Solve.

1 $a^6 \times a^3 =$

2 $(b^2)^4 =$

3 $y^5 \times y^7 =$

4 $(z^5)^5 =$

5 $(a^6)^4 =$

6 $b^5 \times b^2 =$

7 $a^3 \times a^{10} =$

8 $(y^5)^2 =$

9 $b^4 \times b^5 =$

10 $(a^2)^{10} =$

Division property: When dividing powers having the same base, subtract the exponents.

$$a^4 \div a^2 = a^{4-2} = a^2 \qquad \frac{a^4}{a^2} = a^{4-2} = a^2$$

Simplify the expressions.

11 $a^6 \div a^3 =$

12 $\dfrac{b^4}{b^2} =$

13 $\dfrac{y^5}{y^3} =$

14 $z^5 \div z^2 =$

15 $a^{10} \div a^3 =$

16 $\dfrac{y^7}{y^3} =$

17 $z^{125} \div z^{119} =$

18 $\dfrac{b^2}{b^2} =$

19 $a^2 \div a^{-2} =$

20 $a^2 \div y^{-3} =$

Name _____ Date _____

Your Checkbook

RATIONAL NUMBERS

A check register is used to record transactions in a checking account. When you write a check, you are taking money from the account. This amount is written in the Withdrawal column. When you deposit funds, you place money in the account. The balance column records the total positive or negative amount of money in the account.

Complete the Balance column.

	Date	Item	Withdrawal −	Deposit +	Balance =
	9/1				$1,800.50
1.	9/1	Mortgage	$880.00		
2.	9/2	Groceries	$250.00		
3.	9/3	Cash	$80.00		
4.	9/5	Car Payment	$355.86		
5.	9/5	Phone Bill	$75.65		
6.	9/7	Gift		$40.00	
7.	9/8	Water Bill	$72.65		
8.	9/9	Electric Bill	$289.46		
9.	9/9	Pet Food	$35.23		
10.	9/10	Sale of Old Car		$750.00	
11.	9/15	Pay Check		$1,200.00	
12.	9/16	Student Loan	$250.00		
13.	9/16	Groceries	$265.98		
14.	9/17	Car Insurance	$365.00		

Standards-Based Math • 7–8 © 2004 Creative Teaching Press

Be Rational!

RATIONAL NUMBERS

1 $a^6 \times a^3 =$

2 $(b^2)^4 =$

3 $y^5 \times y^7 =$

4 $(z^5)^5 =$

Use mental math to solve.

5 $\dfrac{1}{5}$

6 $\dfrac{15}{50}$

7 $\dfrac{7}{27}$

8 $\dfrac{3}{12}$

9 $\dfrac{4}{33}$

10 $\dfrac{8}{24}$

11 $\dfrac{24}{60}$

12 $\dfrac{4}{18}$

Use <, >, or = to complete.

13 $5\dfrac{1}{2}$ $\bigcirc$ $5.\overline{55}$

14 $\dfrac{22}{66}$ $\bigcirc$ $\dfrac{22}{66}$

15 $^-12 \times 3$ $\bigcirc$ $^-(6^2)$

16 $^-5$ $\bigcirc$ $^-5\dfrac{1}{9}$

Simplify the expressions.

17 $\dfrac{y^7}{y^3} =$

18 $z^{125} \div z^{119} =$

19 $\dfrac{b^2}{b^2} =$

20 $a^2 \div a^{-2} =$

Combinations

PROBABILITY

There are five marbles in a bag: 1 red, 1 orange, 1 green, 1 blue, and 1 yellow. If you pull out two marbles together, how likely are you to pull out a red and an orange marble?

R = red   O = orange   G = green   B = blue   Y = yellow

RO
RG   OG
RB   OB   GB
RY   OY   GY   BY

There are ten different combinations when the marbles are drawn from the bag two at a time. Each is equally likely to occur, so any of the combinations has a $\frac{1}{10}$ probability.

Find the combinations and calculate the probability for each event.

1 Roll two dice. Complete the chart and find the probability of rolling a 1 and a 6.

1, 1
1, 2   2, 2
1, 3   2, 3   3, 3
1, 4   2, 4   3, 4   4, 4
1, 5   2, 5   3, 5
1, 6   2, 6   3, 6

2 Toss four coins in the air. What is the probability that the coins will all land heads up?

3 Three people are needed to work each day at a construction site. There is a group of part-time construction workers who share the job. To keep things fair, every evening they pull names from a hat to determine who will work the next morning. What are the chances Emma, Al, and Ben will work tomorrow?

Al   Ben   Chad   Dan
Emma   Fred   Gigi   Hanna

4 In a new housing development, buyers have a choice of five house models they can build on their lot (A, B, C, D, and E). Itty Bitty Court is a tiny cul–de–sac in the development with space for only three houses and their yards. Assuming that all of the models are equally popular, what are the chances that Itty Bitty court might have Houses B, C, and D?

Standards-Based Math • 7–8 © 2004 Creative Teaching Press

Name _____ Date _____

Permutations

PROBABILITY

Mama Cat has three kittens (A, B, and C). Today they left their box twice to go explore. How many different ways are there for the kittens to leave the box?

ABC	BAC	CAB
ACB	BCA	CBA

There are six different permutations for the three kittens to leave the box. Kitten personality aside, each event is equally likely to occur, so any of the combinations has a $\frac{1}{6}$ probability.

Create a list or picture to find the solution.

1 A game randomly scrambles the letters in TEAM. What are the chances that the new order will spell MEAT?

2 Chris, Alex, Vesna, Jacob, and Jose stand in line for lunch every day. What are the chances that the order today is Alex, Jacob, Jose, Vesna, and then Chris?

3 The following tiles are in a bag:

What are the chances of getting this order?

4 The following tiles are in a bag:

What are the chances of getting this order?

5 The following tiles are in a bag:

What are the chances of getting this order?

Standards-Based Math • 7–8 © 2004 Creative Teaching Press

Name _____ Date _____

Impossible and Certain

PROBABILITY

What is the probability of spinning a 5?

 It is impossible to spin a 5 because the spinner does not contain this number. 5 is not a possible event.

The opposite of an impossible event is a certain event. On a spinner that contains only 5, it is certain that 5 will be the outcome.

Read each situation and write certain or impossible after each.

1 You are playing Add-'Em-Up Domino with your 3-year-old brother using regular 9-dot dominos. The rules are simple. Each player earns the number of points on the open end of each domino he or she places on the board. Your brother is pretty good at matching the dots but is still a little fuzzy with his number identification. He places one domino and announces that he gets "20!" How likely is this?

2 Now your brother is thirsty. You take him to the kitchen and he announces he wants a juice box. All the juice boxes are apple juice. Sticking out his lower lip, he places his hands on his hips and announces authoritatively, "I only want the apple ones!" You laugh. Are you laughing because his choice is certain or impossible?

3 Surprise! You open your closet this morning to find that your beloved cat has given birth to seven kittens! A closer examination reveals that all the kittens are males. Your uncle offers to take three of the kittens to work as mousers on his farm. What is the probability he will choose a boy kitten?

4 You choose a number from the following list: 2, 3, 5, 7, 11, 13, 17, 19. What is the probability that the number will be a composite number?

5

What is the probability that the number will be even?

6

What is the probability that the number will be odd?

7

What is the probability that the number will be 5?

8

What is the probability that the number will be prime?

Standards-Based Math • 7–8 © 2004 Creative Teaching Press

Name _____ Date _____

Calculating Chance

PROBABILITY

$$\text{Probability} = \frac{\text{number of particular events}}{\text{total possible events}}$$

There are 20 marbles in the bag. Three marbles are green. Six marbles are yellow. One marble is black. Five of them are purple. The remaining marbles are purple and gold swirled together. After you pick a marble, you look at the color, note it, and then put the marble back in so that each time you pick there are always 20 marbles.

Read and solve each problem.

1 What is the probability that you will pick a solid yellow marble?

2 What is the probability that you will pick a solid green marble?

3 What is the probability that you will pick a swirled marble?

4 What is the probability that you will pick a solid black marble?

5 What is the probability that you will pick a solid purple marble?

6 What is the probability that you will pick any solid marble?

7 What is the probability that you will pick a marble with gold on it?

8 What is the probability that you will pick a yellow or green marble?

You dump a box of 100 numbered sticks on the floor. What is the probability that . . .

9 you pick up a stick with a multiple of 5? _____

10 you pick up a stick with a multiple of 10? _____

11 you pick up a stick with a multiple of 2? _____

12 you pick up a stick with a multiple of 8? _____

13 you pick up a stick with a multiple of 5 and 8? _____

14 you pick up a stick with a multiple of 2 followed by another multiple of 2? (Assume you return the stick after the first pick.) _____

Answer Key

Natural Numbers and Decimals (page 5)

1. natural
2. decimal
3. natural
4. decimal
5. natural
6. decimal
7. decimal
8. decimal
9. 266,175
10. 184,992
11. 222,075
12. 172,082
13. 277,119
14. 163,548
15. 322,966
16. 147,114
17. 289,488
18. 435,120
19. 2,246,400
20. 1,462,147

Rational and Irrational Decimals (page 6)

Terminating

1/2 = 0.5	1/4 = 0.25
1/5 = 0.2	1/8 = 0.125
1/10 = 0.1	3/4 = 0.75
2/5 = 0.4	3/8 = 0.375
3/10 = 0.3	3/5 = 0.6
5/8 = 0.625	$\sqrt{4}$ = 2.0
$\sqrt{25}$ = 5.0	$\sqrt{9}$ = 3.0

Repeating

1/3 = 0.$\overline{3}$	1/6 = 0.1$\overline{6}$
1/7 = 0.$\overline{142857}$	1/9 = 0.$\overline{1}$
2/3 = 0.$\overline{6}$	5/6 = 0.8$\overline{3}$
2/7 = 0.$\overline{285714}$	5/7 = 0.$\overline{714285}$
2/9 = 0.$\overline{2}$	7/9 = 0.$\overline{7}$
6/7 = 0.$\overline{857142}$	4/9 = 0.$\overline{4}$
8/9 = 0.8	

Irrational

π, $\sqrt{5}$, $\sqrt{3}$, $\sqrt{6}$, $\sqrt{2}$

Rounding Decimals (page 7)

1. 12.3
2. 3.035

3. 7.79
4. 2.1567
5. 3.45
6. 9.0
7. 3.674
8. 315.697
9. 654.15
10. 0.0
11. 1.630
12. 99.0
13. 6.0
14. 4.0

Estimating Decimal Products and Quotients (page 8)

1. 3	11. 4
2. 400	12. 7
3. 40	13. 150
4. 70	14. 4
5. 900	15. 25
6. 16	16. 10
7. 280	17. 9
8. 640	18. 3
9. 275	19. 3
10. 5	20. 10
	21. 3

Multiplying Decimals (page 9)

1. 45
2. 4.829724
3. 12.15
4. 7.2
5. 0.001675
6. 93.942
7. 16.105
8. 0.0024
9. 113.832
10. 0.1
11. 3,269.6024
12. 35.1384
13. 28,340
14. 6
15. 616.4
16. 20
17. 0.27
18. 53.508
19. 2.7775
20. 0.5

Dividing Decimals (page 10)

1. 2
2. 30
3. 100
4. 500
5. 9,000
6. 69
7. 2
8. 25.2
9. 54.32
10. 100,000
11. 600
12. 2,810
13. 66,054.3
14. 200,000
15. 4,101
16. 5,000
17. 32
18. 4.$\overline{54}$
19. 600
20. 15.825

Exponents (page 11)

1. $5 \times 5 \times 5 = 125$
2. $2 \times 2 \times 2 \times 2 = 16$
3. $3 \times 3 \times 3 \times 3 = 81$
4. $6 \times 6 = 36$
5. $8 \times 8 \times 8 = 512$
6. $9 \times 9 = 81$
7. $5 \times 5 \times 5 \times 5 = 625$
8. $7 \times 7 \times 7 = 343$
9. $4 \times 4 \times 4 = 64$
10. $10 \times 10 \times 10 = 1,000$
11. $12 \times 12 = 144$
12. $11 \times 11 \times 11 = 1,331$
13. $1 \times 1 \times 1 \times 1 \times 1 \times 1 \times 1 \times 1 \times 1 \times 1 \times 1 \times 1 = 1$
14. $4 \times 4 \times 4 \times 4 = 256$
15. $7 \times 7 \times 7 \times 7 \times 7 = 16,807$
16. $5 \times 5 \times 5 \times 5 \times 5 = 3,125$
17. $2 \times 2 \times 2 \times 2 \times 2 \times 2 \times 2 \times 2 = 256$
18. $6 \times 6 \times 6 = 216$
19. $9 \times 9 \times 9 \times 9 = 6,561$
20. $3 \times 3 \times 3 \times 3 \times 3 = 243$
21. $6 \times 6 \times 6 \times 6 = 1,296$

Exponential Notation (page 12)

1. 1,000
2. 100,000,000
3. 10,000,000
4. 720,000
5. 60,000,000
6. 0.00087
7. 9,000,000,000
8. 0.000000052
9. 0.000001204
10. 0.0000006
11. 4,200,000,000,000
12. 11,100,000,000
13. 3,879,000,000
14. 0.000068
15. 9,800,000,000
16. 960,000,000,000,000
17. 0.00000009873
18. 0.00009

Divisibility (page 13)

1. Yes: 2, 3, 4, 6, 8, 9; No: 5, 10
2. Yes: 2, 3, 4, 6, 8, 9; No: 5, 10
3. Yes: 2, 4, 5, 8, 10; No: 3, 6, 9
4. Yes: all
5. Yes: 2, 3, 4, 6, 8, 9; No: 5, 10
6. Yes: 3, 9; No: 2, 4, 5, 6, 8, 10
7. Yes: 2, 4, 5, 8, 10; No: 3, 6, 9
8. Yes: none; No: all
9. Yes: 2, 4, 8; No: 3, 5, 6, 9, 10
10. Yes: 2; No: 3, 4, 5, 6, 8, 9, 10
11. Yes: 2, 4; No: 3, 5, 6, 8, 9, 10
12. Yes: 3, 5; No: 2, 4, 6, 8, 9, 10
13. Yes: 3; No: 2, 4, 5, 6, 8, 9, 10
14. Yes: 2, 3, 6; No: 4, 5, 8, 9, 10
15. Yes: 2; No: 3, 4, 5, 6, 8, 9, 10
16. Yes: 2; No: 3, 4, 5, 6, 8, 9, 10
17. Yes: none; No: all
18. Yes: 3; No: 2,4,5,6,8,9,10

Factors, Primes, and Composites (page 14)

1. prime
2. composite; 2, 4, 8
3. composite; 3, 13
4. composite; 2, 3, 6, 7, 14, 21
5. prime
6. composite; 2, 3, 4, 6
7. 1, ②, ③, 6, ⑦, 14, 21, 42
8. 1, ⑦, ⑪, 77
9. 1, ②, ③, ⑤, 6, ⑦, 10, 14, 15, 21, 30, 35, 42, 70, 105, 210
10. 1, ②, ③, 6, ⑦, 9, 14, 18, 21, 42, 63, 126
11. 1, ②, ⑤, 10, 25, 50
12. 1, ②, 4, ⑤, 8, 10, 20, 25, 40, 50, 100, 200
13. 1, ⑪, ⑬, 143

14. 1, 2, 3, 4, 6, 9, 12, 18, 36
15. 1, 2, 4, 41, 82, 164
16. 1, 2, 61, 122
17. 1, 2, 13, 26
18. 1, 2, 4, 8,16, 32, 64
19. 1, 2, 3, 4, 5, 6, 8, 9, 10, 12, 15, 18, 20, 24, 30, 36, 40, 45, 60, 72, 90, 120, 180, 360
20. 1, 2, 3, 6, 9, 26, 39, 78, 117, 234
21. 1, 2, 3, 4, 6, 8, 12, 16, 24, 48

Prime Factorization (page 15)

1. 2 × 3 × 11
2. 5 × 5 × 13
3. prime
4. 41 × 11
5. 2 × 13 × 19
6. prime
7. 2 × 337
8. 3 × 3 × 11
9. 5 × 5 × 3 × 3
10. 3 × 3 × 41
11. 2 × 2 × 2 × 23
12. 2 × 2 × 2 × 2 × 13
13. 17 × 29
14. 2 × 2 × 37
15. prime
16. 2 × 2 × 3 × 11
17. 5 × 5 × 3
18. 2 × 2 × 2 × 2 × 2 × 2 × 2
19. 23 × 23
20. 2 × 2 × 2 × 2 × 3
21. 2 × 2 × 17

Least Common Multiple (page 16)

1. 40	**7.** 9
2. 35	**8.** 12
3. 24	**9.** 24
4. 36	**10.** 6
5. 60	**11.** 72
6. 48	**12.** 84

Greatest Common Factor (page 17)

1. 12	**6.** 4
2. 9	**7.** 5
3. 7	**8.** 31
4. 8	**9.** 52
5. 15	**10.** 13

Simplifying Fractions (page 18)

1. 2/21	**11.** 2/3
2. 3/5	**12.** 9/16
3. 2/5	**13.** 1/3
4. 3/4	**14.** 1/8
5. 3/4	**15.** 1/3
6. 3/4	**16.** 1/5
7. 1/3	**17.** 2/3
8. 2/3	**18.** 5/11
9. 1/6	**19.** 1/3
10. 5/8	**20.** 1/10

Functions (page 19)

1. × 5
2. + 14
3. ÷ 3
4. × 3
5. – 7
6. ÷ 6
7. × 12
8. – 2
9. + 8
10. × 6
11. + 31
12. ÷ 8
13. × 9
14. – 6
15. ÷ 2
16. – 1/4
17. + 1/2
18. ÷ 4 or × 0.25

Comparing Fractions (page 20)

1. <	11. =
2. >	12. <
3. >	13. >
4. =	14. =
5. <	15. >
6. >	16. >
7. >	17. <
8. <	18. <
9. =	19. =
10. >	20. >

Improper Fraction to Mixed Number (page 21)

1. 1 2/3	**13.** 4 2/5
2. 3 3/4	**14.** 2 1/6
3. 3 1/9	**15.** 4 4/5
4. 2 3/7	**16.** 4 5/11
5. 6 1/2	**17.** 3 5/8
6. 6 4/5	**18.** 5 1/6
7. 3 6/11	**19.** 8 1/6
8. 4 2/3	**20.** 11 1/5
9. 5 2/3	**21.** 8 3/5
10. 1 11/19	**22.** 74
11. 6 13/19	**23.** 5 2/15
12. 5 3/4	**24.** 25 1/5

Mixed Number to Improper Fraction (page 22)

1. 54/7
2. 45/11
3. 57/5
4. 74/9
5. 63/10
6. 66/8 or 33/4
7. 77/6
8. 75/6 or 25/2
9. 78/9 or 26/3
10. 73/8
11. 76/12 or 19/3
12. 64/5
13. 67/8
14. 55/6
15. 46/5
16. 56/5
17. 79/11
18. 65/6
19. 31/8
20. 38/3
21. 57/6
22. 29/2
23. 83/4
24. 127/8

Adding and Subtracting Fractions (page 23)

1. 11/12
2. 1/6
3. 1/18
4. 4/5
5. 4/5
6. 19/27
7. 11/12
8. 1/12

9. 10/27
10. 4/11
11. 23/24
12. 7/24
13. 13/15
14. 1/3
15. 1/30
16. 1/2
17. 7/8
18. 3/40
19. 5/12
20. 1 1/24
21. 11/12
22. 7/90
23. 23/24
24. 1/12

Adding Mixed Numbers (page 24)

1. 6 7/8
2. 8 4/9
3. 15 14/15
4. 6 31/36
5. 12 1/15
6. 12 1/15
7. 13 13/20
8. 18 5/12
9. 27 1/6
10. 7 31/40
11. 13 2/7
12. 9 1/9
13. 13 3/130
14. 9 19/36
15. 11 143/144
16. 13 7/40
17. 8 55/63
18. 15 27/35

Subtracting Mixed Numbers (page 25)

1. 2 3/8	**10.** 1 3/8
2. 3 1/15	**11.** 2 1/6
3. 2 1/9	**12.** 5 1/6
4. 2 1/12	**13.** 1 7/20
5. 5 11/48	**14.** 3 1/20
6. 3 13/36	**15.** 5 7/36
7. 2 1/2	**16.** 2 5/8
8. 3 5/18	**17.** 4 2/45
9. 6 3/14	**18.** 1

Factors and Fractions (page 26)

1. $\sqrt{6}$,
2. $0.\overline{55}$
3. 2.16
4. 16
5. 4
6. 113.832
7. 2
8. 16.105
9. $1,666.\overline{6}$
10. 7.2
11. 216
12. 0.00087
13. 14 11/80
14. 2 3/8

Fractions to Percents (page 27)

1. 0.33; 33%
2. 0.28; 28%
3. 0.33; 33%
4. 0.56; 56%
5. 0.19; 19%
6. 0.21; 21%
7. 0.26; 26%
8. 0.12; 12%
9. 0.33; 33%
10. 0.09; 9%
11. 70%
12. 18%
13. 90%
14. 90%
15. 65%
16. 74%
17. 53%
18. 33%
19. 43%
20. 62%

Multiplying Fractions (page 28)

1. 3/21
2. 21/32
3. 2/15
4. 1/4
5. 10/27
6. 1/7
7. 3/8
8. 2/7
9. 4/21
10. 4/9
11. 2/7
12. 3/20
13. 5/49
14. 3/25
15. 15/56
16. 1/14
17. 1/8
18. 1/4
19. 1/7
20. 1/5
21. 1/2
22. 5/21
23. 12/175
24. 1/32

Multiplying Mixed Numbers (page 29)

1. 23 7/24
2. 12 2/3
3. 15 5/12
4. 22 27/32
5. 6 8/9
6. 26 1/4
7. 35 7/15
8. 8 6/7
9. 7 1/5
10. 6 1/2
11. 2 1/4
12. 1 1/8
13. 18
14. 4 1/4
15. 15
16. 1 3/7
17. 15
18. 12 4/5
19. 21
20. 2
21. 2/9

Dividing Fractions (page 30)

1. 2 5/8
2. 2/9
3. 5/7
4. 5/6
5. 27/35
6. 21/32
7. 1 2/3
8. 7/10
9. 3/5
10. 16/21
11. 1 2/3
12. 3/4
13. 1 2/7
14. 7/24
15. 25/36
16. 12
17. 21
18. 2 2/5
19. 3
20. 1/2
21. 16

Dividing Mixed Numbers (page 31)

1. 3 1/2
2. 13 1/3
3. 1 1/3
4. 2 2/3
5. 1/8
6. 4 1/2
7. 3/8

8. 14
9. 2 2/3
10. 4 1/2
11. 8 4/5
12. 4
13. 33
14. 9 2/5
15. 6
16. 7 1/5
17. 1/5
18. 3/4
19. 10 1/2
20. 9
21. 5

Multistep Problems (page 32)

1. 26 days with 3 lb. remaining
2. $230.00
3. paying the full-time rate saves you $15.00
4. $3,640.00
5. $3,087.70

Along the Way (page 33)

Acoss
2. natural
5. factor
7. denominator
8. least common multiple
9. decimals
10. irrational
11. numerator

Down
1. greatest common factor
3. rational
4. prime
6. composite

Altogether (page 34)

1. 33%
2. 26%
3. 70%
4. 43%
5. 21/32
6. 1/7
7. 3/25
8. 1/4
9. 8/105
10. 1/32
11. 7 1/5
12. 12 4/5
13. 2/9
14. 2/9
15. 5/7
16. 21/32
17. 25/36
18. 21
19. 2 2/3
20. 33
21. 3/4
22. 14

Variables (page 35)

1. 3 + n = 8 books
2. n + 200 = 1,000 miles
3. 2 + x = 7
4. x + 12 = 48
5. 2 + x = 8
6. 1/3x = 1,000 lb.

Associative and Commutative Properties (page 36)

$\times$	2	4	$\frac{1}{2}$	-3	5
$\frac{1}{3}$	$\frac{2}{3}$	$1\frac{1}{3}$	$\frac{1}{6}$	-1	$1\frac{2}{3}$
6	12	24	3	-18	30
$\frac{3}{5}$	$1\frac{1}{5}$	$2\frac{2}{5}$	$\frac{3}{10}$	$-1\frac{4}{5}$	3
5	10	20	$2\frac{1}{2}$	-15	25
8	16	34	4	-24	40

2 × 1/3; 1/3 × 2
2 × 6; 6 × 2
2 × 3/5; 3/5 × 2
2 × 5; 5 × 2
2 × 8; 8 × 2
4 × 1/3; 1/3 × 4
4 × 6; 6 × 4
4 × 3/5; 3/5 × 4
4 × 5; 5 × 4
4 × 8; 8 × 4
1/2 × 1/3; 1/3 × 1/2
1/2 × 6; 6 × 1/2
1/2 × 3/5; 3/5 × 1/2
1/2 × 5; 5 × 1/2
1/2 × 8; 8 × 1/2
-3 × 1/3; 1/3 × -3
-3 × 6; 6 × -3
-3 × 3/5; 3/5 × -3
-3 × 5; 5 × -3
-3 × 8; 8 × -3
5 × 1/3; 1/3 × 5
5 × 6; 6 × 5
5 × 3/5; 3/5 × 5
5 × 5; 5 × 5
5 × 8; 8 × 5

1. 48
2. 19
3. 60
4. 23
5. 42
6. 60
7. 42
8. 109
9. 44
10. 60
11. 969
12. 48
13. 379
14. 108

Distributive Property (page 37)

1. 21	**15.** −60
2. -32	**16.** 80
3. 22	**17.** 135
4. 10	**18.** 378
5. 56	**19.** 64
6. 45	**20.** 486
7. -32	**21.** 56
8. 33	**22.** 54
9. 28	**23.** -144
10. 207	**24.** 45
11. 50	**25.** 48
12. 72	**26.** 108
13. 75	**27.** 10
14. 96	

Order of Operations (page 38)

1. 19	**8.** 61
2. 4	**9.** 50
3. 64	**10.** 7
4. 5	**11.** 4
5. 8	**12.** 56
6. 8	**13.** 3
7. 125	**14.** 12

Inverse Operations (page 39)

1. E	**7.** B
2. G	**8.** D
3. J	**9.** L
4. F	**10.** I
5. C	**11.** A
6. K	**12.** H

Inverse Operations: + (Page 40)

1. x = 8	**11.** x = 20
2. a= 27	**12.** c = 0
3. y = 13	**13.** x = 5
4. z = 42	**14.** x = 40
5. x = 10	**15.** n = 3
6. a = 85	**16.** r = ⁻5
7. x = 85	**17.** n = 21
8. a = 18	**18.** t = 20
9. y = 59	**19.** x = 48
10. 36	**20.** x = 3/4

Inverse Operations: – (page 41)

1. x = 16	**11.** x = 4
2. a = 30	**12.** c = 4
3. y = 25	**13.** a = 72
4. z = 95	**14.** x = 74
5. x = 85	**15.** x = 10 3/8
6. a = 145	**16.** c = 109
7. x = 115	**17.** y = 14
8. a = 33	**18.** b = 16 11/12
9. y = 22	**19.** c = 855
10. b = 6.4	**20.** b = 138

Inverse Operations: × (page 42)

1. 5	**11.** 100
2. 10	**12.** 5
3. 9	**13.** 7
4. 20	**14.** 7
5. 5	**15.** 7
6. 6	**16.** 56
7. 7	**17.** 4
8. 7	**18.** 2
9. 9	**19.** 7.5
10. 15	**20.** 13

Inverse Operations: ÷ (page 43)

1. 42	**11.** 25
2. 48	**12.** 55
3. 48	**13.** 36
4. 45	**14.** 35
5. 20	**15.** 30
6. 120	**16.** 120
7. 27	**17.** 135
8. 75	**18.** 24
9. 6	**19.** 165
10. 88	**20.** 100

Variables 2 (page 44)

1. 5	**11.** 9
2. 2	**12.** 49
3. 12	**13.** ⁻3
4. 6	**14.** 3
5. 7	**15.** ⁻12
6. 1	**16.** ⁻13
7. 3	**17.** ⁻9
8. 20	**18.** ⁻7
9. 4	**19.** 3
10. 100	**20.** 9

Inequalities (page 45)

1. < 3
2. > −144
3. <−1 1/3
4. > 61
5. < 194
6. > 1
7. < 2/3
8. < −55
9. < 1
10. < −180
11. > 1/8
12. > 16
13. > −2
14. ≥ 1
15. ≤ 44
16. ≥ 3 1/9
17. ≤ 4
18. > −1.8
19. ≥ −5
20. < −4

Graphing Inequalities (page 46)

1.
2.
3.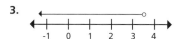
4.
5.
6.
7.
8.
9.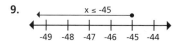
10.

Lines and Line Segments (page 47)

1. $\overleftrightarrow{GH}$
2. $\overline{KL}$
3. $\overleftrightarrow{EF}$ intersects $\overleftrightarrow{GH}$
4. $\overleftrightarrow{AT} \perp \overrightarrow{MY}$
5. $\overleftrightarrow{CT} \parallel \overleftrightarrow{DG}$
6. $\overrightarrow{ME} \perp \overrightarrow{WE}$
7. $\overleftrightarrow{ET} \perp \overleftrightarrow{VU}$
8. $\overline{RS} \parallel \overleftrightarrow{TU}$

Angles (page 48)

1. B
2. F
3. D
4. A
5. E
6. C

Answers will vary. Possible answers include:

7.

8.

9.

10. 90°
11. 130°
12. 40°

Angle Measurement (page 49)

1. 60°
2. 47°
3. 47°
4. 15°
5. 45°
6. 142°
7. 59°
8. 75°
9. 48°
10. 43°
11. 48°
12. 140°

Geometry Terms (page 50)

Across
2. plane
3. point
4. parallel
5. edge
6. diameter
8. chord
9. solid
11. intersecting
14. face

Down
1. midpoint
3. perpendicular
7. radius
10. angle
12. tangent
13. ray

Polygons (page 51)

1. triangle
2. heptagon
3. quadrilateral
4. nonagon
5. pentagon
6. hexagon
7. quadrilateral
8. dodecagon
9. triangle
10. octagon
11. quadrilateral
12. decagon
13. triangle
14. decagon
15. quadrilateral

Similar and Congruent (page 52)

1. similar
2. congruent
3. congruent
4. similar
5. congruent
6. similar
7. similar
8. congruent
9. congruent
10. congruent
11. similar
12. similar

Similar Triangles (page 53)

1.

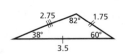

2.

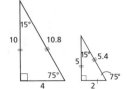

3.

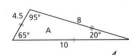

4.

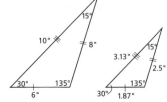

Similar Figures (page 54)

1.

1.5 1.5
3   3
3

2.

3.6   3.6
8   8
4.8

3.

4.5   3   4.5
4.5   3   4.5

4.

12.9
9   16
9.6
14.1

5.

7.5
8.25
4.5
6

6.

15
6   6
15

7.

10
5
10   5   5   10
5
10

8.

15
15   13.5
4.5   5
5

Parts of a Circle (page 55)

1. C
2. E
3. D
4. A
5. F
6. B

Answers will vary. Possible answers include:

118

Geometric Solids (page 56)

1. cylinder
2. pyramid, triangle
3. prism, hexagon
4. sphere
5. cone
6. prism, triangle
7. cylinder
8. pyramid, rectangle
9. pyramid, pentagon
10. prism, octagon
11. prism, rectangle
12. They both have one base and come to a point.

Spatial Relationships (page 57)

1. I	7. D
2. F	8. A
3. K	9. E
4. C	10. J
5. G	11. H
6. B	

Angles, Circles, and Solids (page 58)

1. line
2. line segment
3. tangent
4. pyramid
5. prism
6. acute angle
7. obtuse angle
8. right angle
9. complementary angle
10. cone
11. cylinder
12. sphere

The X, Y Axis (page 59)

1. (–8, 9)	8. (1, –2)
2. (–8, 5)	9. (–6, –1)
3. (–5, 8)	10. (–4, –4)
4. (3, –5)	11. (–4, –1)
5. (7, –4)	12. (1, 4)
6. (7, 1)	13. (3, 4)
7. (3, 2)	14. (2, 7)

Using a Cartesian Plane (page 60)

1. (–1, 3)	6. (6, 0)
2. (–1, 4)	7. (–3, 4)
3. (–2, 3)	8. (–6, 11)
4. (5, 9)	
5. (–8, –9)	

Reflections Across the X-Axis (page 61)

1. Reflection (2, 8)
(2, 2)
(4, 2)
(4, 6)
(7, 6)
(7, 8)

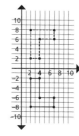

2. Reflection (1, –3)
(1, –8)
(5, –8)
(5, –6)
(4, –6)
(4, –5)
(5, –5)
(5, –3)

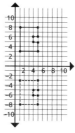

3. Reflection (–3, –3)
(–5, –6)
(–3, –9)
(–5, –9)
(–6, –7)
(–7, –9)
(–9, –9)
(–7, –6)
(–9, –3)
(–7, –3)
(–6, –5)
(–5, –3)

4. Reflection (–6, 2)
(–4, 4)
(–6, 5)
(–4, 8)
(–8, 8)
(–10, 5)
(–8, 4)
(–10, 2)

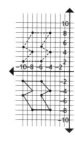

Reflections Across the Y-Axis (page 62)

1. Reflection (2, 6)

(2, 8)
(5, 8)
(5, 7)
(7, 7)
(7, 4)
(5, 4)
(5, 6)

2. Reflection (3, 3)

(3, 6)
(4, 5)
(4, 8)
(7, 7)
(7, 5)
(6, 6)
(6, 3)
(7, 2)
(4, 2)

3. Reflection (−2, 4)

(−2, 6)
(−4, 9)
(−5, 7)
(−8, 9)
(−8, 6)
(−5, 6)
(−6, 4)

4. Reflection (−4, 5)

(−6, 7)
(−9, 7)
(−11, 5)
(−8, 5)
(−10, −2)
(−5, −2)
(−7, 5)

Translations Along the X-Axis (page 63)

1. Translation (−2, 4)

(1, 8)
(2, 4)

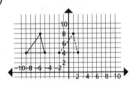

2. Translation (0, 5)

(1, 10)
(3, 8)
(3, 5)

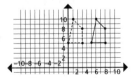

3. Translation (−1, −9)

(3, −3)
(4, −6)
(2, −9)

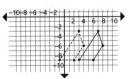

4. Translation (2, 2)

(1, 4)
(2, 6)
(1, 8)
(2, 10)
(4, 10)
(3, 8)
(4, 6)
(3, 4)
(4, 2)

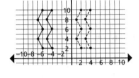

Translations Along the Y-Axis (page 64)

1. Translation (2, 0)

(2, 2)
(8, 0)
(7, −1)
(4, −1)

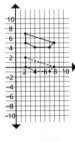

2. Translation (−10, 3)

(−9, −1)
(−7, 0)
(−5, 2)
(−8, 2)

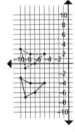

3. Translation (3, −5)

(3, −3)
(5, −2)
(7, −3)
(9, −2)
(9, −5)

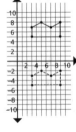

4. Translation (−10, 6)

(−7, 6)
(−5, 4)
(−3, 6)
(−3, 1)
(−5, 3)
(−7, 1)
(−10, 1)

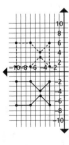

Oblique Translations (page 65)

1. Translation (7, –1)
(7, 0)
(11, 0)
(11, –1)

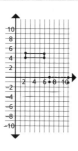

2. Translation (–6, 9)
(–6, 13)
(–4, 11)
(–4, 8)

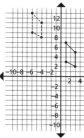

3. Translation (9, –5)
(6, –7)
(9, –8)
(12, –7)

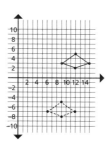

4. Translation (4, 6)
(6, 4)
(5, 3)
(6, 1)
(4, 1)
(3, 3)
(4, 4)
(2, 6)

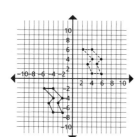

Transformations (page 66)

1. reflection across Y-axis
2. reflection across X-axis
3. reflection across Y-axis
4. oblique translation
5. translation along Y-axis
6. oblique translation

Perimeter (page 67)

1. 16"
2. 22 ft
3. 48 m
4. 45 cm
5. 40

6. 32 km
7. 60
8. 36
9. 80
10. 28.5
11. 31
12. 16

Area (page 68)

1. 6.25
2. 11.4
3. 1,323 sq. in. or 110.25 sq. ft.
4. 4 sq. cm
5. 1, 200
6. 80 sq. km
7. 51.84
8. 14.95
9. 63
10. 175 sq. m
11. 1,600
12. 320

Area of a Triangle and Trapezoid (page 69)

1. 75
2. 500
3. 120
4. 1.2 sq. m
5. 8
6. 94.5
7. 72
8. 36.75 sq. in.
9. 7, 700
10. 56.25
11. 510
12. 112 sq. mi.

Circumference (page 70)

1. 131.88 cm
2. 94.2
3. 18.84
4. 263.76
5. 508.68 in.
6. 75.36
7. 628
8. 125.6
9. 50.24 m
10. 20.096 cm
11. 36.424
12. 1.319

Area of Circles (page 71)

1. 1,384.74
2. 706.5 sq. cm
3. 28.26
4. 5,538.96 sq. cm
5. 20,601.54
6. 452.16 sq. m
7. 31,400 sq. mi.
8. 1,256
9. 200.96
10. 32.1536
11. 105.6296 sq. ft.
12. 0.138474

Area and Circumference of Circles (page 72)

1. A = 50.24; C = 25.12
2. A = 200.96; C = 50.24
3. A = 113.04; C = 37.68
4. A = 615.44; C = 87.92
5. A = 78.5; C = 31.4
6. A = 907.46; C = 106.76
7. A = 1,256; C = 125.6
8. A = 1,962.5; C = 157
9. A = 153.86; C = 43.96

Points, Shapes, and Solids (page 73)

1. x = 2
2. x = 7
3. x ≈ 4.127
4. x ≈ 9.3
5. (2, –4)
6. (–2, 4)
7. (6, 2)
8. (–3, –5)
9. A = 16; P = 16
10. A = 100; P = 58

Ratios (page 74)

1. There are 4 pieces of candy for each student.
2. About 10" of rain fell per month.
3. Someone can type 70 words per minute.
4. The car uses 1 gal. of gas for every 22 miles traveled.
5. There are 8 seeds for every pot.
6. The object traveled about 7 m per second.
7. There are 50 stars on each flag.
8. An average of one hour of each day is spent knitting.
9. There are five pencils for each student.
10. Ratio: 5:100 or 1:20

11. Ratio: 1:3
12. Ratio: 2":12" or 1": 6"

Rates (page 75)

1. 1,200 m/3 days or 400 m/day
2. $260/20 days or $13/day
3. 300 flowers/20 tables or 15 flowers/table
4. 480 beads/12 necklaces or 40 beads/necklace
5. $810/9 hours or $90/hr
6. 2,500 fish/50 tanks or 50 fish/tank
7. 14.5 miles/hr.
8. 86 km/hour
9. 3 yards/dress
10. $25/class
11. 4 mi./day
12. 4 eggs/omelet

Proportions (page 76)

1. a = 4
2. d = 12
3. c = 1
4. x = 18
5. b = 4
6. d = 20
7. b = 8
8. d = 3
9. a = 4
10. d = 8
11. x = 45
12. x = 35
13. x = 63
14. x = 3
15. n = 9

Cross Multiplication (page 77)

1. d = 4.5
2. b = 10
3. b = 4
4. c = 8
5. a = 2
6. a = 1
7. d = 15
8. d = 24
9. x = 75
10. a = 48
11. x = 5
12. a = 6
13. x = 2.$\overline{66}$
14. x = 3
15. x = 9

Proportions 2 (page 78)

1. 4
2. 8
3. 12
4. 9
5. 81
6. 20
7. 2
8. 24
9. 2.5
10. 8
11. 5
12. 5
13. 16
14. 30
15. 9
16. 7.5
17. 21
18. 5
19. 15
20. 12

Unit Price (page 79)

1. 21¢
2. $3.00
3. $4.25
4. $1.29
5. $23.60
6. 38¢
7. 29¢
8. $2.99
9. $1.75
10. $1.19/lb; $3.50/lb; apples
11. $1.66/pair; $1.30/pair; 12 pairs
12. $1.79/lb; $1.80/lb; 5lb.
13. $2.15/gallon; $2.16/ gallon; 8 gallons
14. $23.60/gallon; $3.99/gallon; 1 gallon
15. $4.87/oz.; $6.49/oz.; 8 oz.

Ratio and Proportion (page 80)

1. 28/1
2. 13/21
3. 4:5
4. 1:11
5. 1:4
6. 2:3
7. 2
8. 24
9. 2.5
10. 8
11. 5
12. 5
13. 15
14. 12
15. 250 mi./day
16. $9.75

Percent (page 81)

1. 20%
2. 25%
3. 30%
4. 40%
5. 5%
6. 5%
7. 80%
8. 80%
9. 10%
10. 100%
11. 62.5%
12. 40%
13. 15%
14. 8%
15. 6%
16. 7%
17. 11%
18. 88%
19. 63%
20. 28%

Fractions, Decimals, and Percents (page 82)

1. 33%
2. 28%
3. 33%
4. 56%
5. 19%
6. 21%
7. 26%
8. 12%
9. 33%
10. 9%
11. 33%
12. 228%
13. 70%
14. 20%
15. 10%
16. 60%
17. 66%
18. 88%

More Fractions, Decimals, and Percents (page 83)

	Percent	Decimal	Fraction		Percent	Decimal	Fraction
1.	45%	0.45	9/20	15.	50%	0.5	1/2
2.	210%	2.1	2 1/10	16.	44%	0.44	11/25
3.	0.4%	0.004	4/1000	17.	44%	0.$\overline{44}$	4/9
4.	12.3%	0.123	123/1000	18.	75%	0.75	3/4
5.	30%	0.3	3/10	19.	9%	0.09	9/100
6.	33 ⅓%	0.$\overline{33}$	1/3	20.	80%	0.8	4/5
7.	12.5%	0.125	1/8	21.	1.5%	0.015	3/200
8.	1%	0.01	1/100	22.	78%	0.78	39/50
9.	9%	0.$\overline{09}$	1/11	23.	28%	0.28	7/25
10.	14%	0.14	7/50	24.	95%	0.95	19/20
11.	87.5%	0.875	7/8	25.	32%	0.32	8/25
12.	66 ⅔%	0.$\overline{66}$	2/3	26.	18%	0.18	9/50
13.	16 ⅔%	0.1$\overline{6}$	1/6	27.	98%	0.98	49/50
14.	25%	0.25	1/4	28.	20%	0.2	1/5

Percent of a Number (page 84)

1. 4
2. 245
3. 750
4. 5
5. 16
6. 24 mL
7. 5 mL
8. 138
9. 23
10. 95
11. 15
12. 108
13. 100
14. 20
15. $8.00
16. 15

Data Analysis (page 85)

1. 4-H, ABC Daycare, Boy Scouts, Girl Scouts, Larson's Tire, Acme Hardware, Central Farm Co-op, Central High Marching Band, Bob's Feed and Grain
2. average = 4; mode = 0 or 2. Kylie knows that the 4-H float is skewing the average and making it seem that there are about four animals in each float. The mode of 0 or 2 tells the judges that most floats have very few animals on them.
3. average = 1.93 Tons. The parade committee does not have to pay the special tax.
4. 20 people

Mean, Median, Mode, and Range (page 86)

1. 10; 9; 9 and 12; 9
2. 0.455; 0.445; all values; 0.51
3. 84.75; 90; 64; 77
4. 25; 23; 30; 21
5. 17,110; 17,423; all values; 5,877
6. $62.00; $44.00; $44.00; $136.00
7. 79%; 78.5%; 73%; 32%
8. 78; 77; 107; 76

Reading Tables (page 87)

1. 6 children
2. brown
3. Peterson, A. and Peterson, B.
4. time born; first to last
5. 6 lb. 13.5 oz
6. 19.46 in.
7. 3 lb.
8. neither

Reading Line Graphs (page 88)

1. 29 in.; ~2 in.; 15 in.
2. between 2 and 4 months; 22 and 24 months
3. The federal minimum wage; no; the measurement of money (dollars? cents?)
4. 2 hr.; riding up between 20 and 60 minutes, riding down between 60 and 80 minutes

Creating a Circle Graph (page 89)

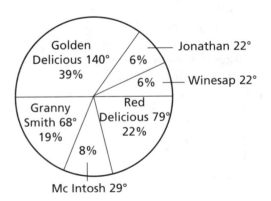

Scatter Plots (page 90)

Answers will vary. Possible answers include:

1.

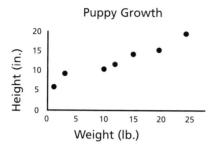

Puppy Growth

2.

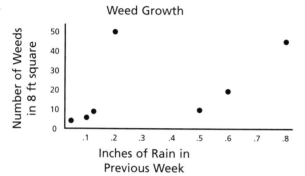

Weed Growth

3.

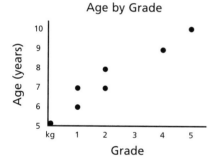

Age by Grade

4.

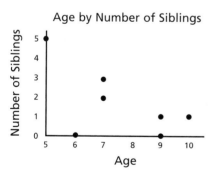

Age by Number of Siblings

5. Age by Number of Siblings

Reading Graphs (page 91)

1. Plainview, MN

2. Corona, CA

3. Plainview: August to October. Corona: March to April

4. 12th grade

5. 7%

6. 34.6%

7. 70 miles

8. no

9. 350 miles

The Bottom Line (page 92)

1. Sunday night

2. 2 hours

3. 1.94 hours

4. The higher the amount of TV watched, the lower the test score.

5. 10

6. It's evenly distributed with five data points above six hours and five data points at 4 hours or less.

Defining Integers (page 93)

1. 7, -9

2. 12, -12

3. 0, -8

4. -2, -13

5. 1" + 2" + - 4" = -1"

6. $30 + ⁻$10 + ⁻$15 + $10 = $15

7. –$1,000 + $200 = –$800

8. –$5.00 + $12.00 = $7.00

Absolute Value (page 94)

1. 4

2. 5

3. 4

4. 4

5. –7

6. –7

7. –6

8. 4

9. 24

10. 16

11. 20

12. ⁻2

13. 10

14. 26

15. 6

16. –44

17. −79
18. 24
19. −1/4
20. 1

Adding Integers (page 95)

1. −4
2. 7
3. −59
4. −14
5. 12
6. −14
7. 27
8. 13
9. −26
10. −48
11. 23
12. −9
13. −54
14. 3
15. 12
16. 1

Subtracting Integers (page 96)

1. −72 + (−70) = −142
2. 52 + 46 = 98
3. 90 + (−49) = 41
4. 7 + 95 = 102
5. −55 + 59 = 4
6. −47 + (−20) = − 67
7. −13
8. 56
9. −145
10. −57
11. −3
12. 24
13. −2
14. 141
15. −49
16. 30
17. −9
18. −76
19. 50
20. 87

Exploring Linear Equations (page 97)

1. x + 1, y + 1
2. x − 1, y + 3
3. x + 4, y − 2
4. 2x, y + 1
5. x + 1, 3y
6. 2x + 1, y

Positive and Negative (page 98)

1. −7
2. 0
3. 4
4. 24
5. 16
6. 20
7. 2
8. 27
9. 13
10. −26
11. −48
12. −3
13. −172
14. 141
15. 2x, y + 1
16. x + 1, 3y

Terminating Decimals (page 99)

1. 0.2
2. 0.25
3. 0.3
4. 0.4
5. 0.05
6. 0.05
7. 0.8
8. 0.8
9. 0.1
10. 1
11. 0.625
12. 0.4
13. 0.15
14. 0.08
15. 0.525
16. 0.2
17. 0.4375
18. 0.45
19. 0.04
20. 0.0975

Repeating Decimals (page 100)

1. $0.\overline{3}$
2. $0.2\overline{7}$
3. $0.\overline{3}$
4. $0.\overline{5}$
5. $0.\overline{190476}$
6. $0.208\overline{3}$
7. $0.\overline{259}$
8. $0.\overline{12}$
9. $0.\overline{3}$
10. $0.\overline{2}$
11. $0.\overline{5}$
12. $0.\overline{6}$
13. $0.\overline{2}$
14. $0.1\overline{6}$
15. $0.1\overline{6}$
16. $0.\overline{5}$
17. $0.\overline{3}$
18. $0.58\overline{3}$
19. $0.\overline{6}$
20. $0.0\overline{925}$

Comparing Rational Numbers (page 101)

1. <
2. =
3. =
4. >
5. <
6. >
7. >
8. >
9. <
10. =
11. =
12. >, >
13. >
14. <
15. >
16. =
17. =
18. <

The Density Property (page 102)

These problems have many correct answers. Accept all reasonable responses.

Scientific Notation (page 103)

1. a^9
2. b^8
3. y^{12}
4. 2^{25}
5. a^{24}
6. b^7
7. a^{13}
8. y^{10}
9. b^9
10. a^{20}
11. a^3
12. b^2
13. y^2
14. z^3
15. a^7
16. y^4
17. z^6
18. b^0
19. a^4
20. a^5

Your Checkbook (page 104)

1. 920.5
2. $670.5
3. $590.5
4. $234.64
5. $158.99
6. $198.99
7. $126.34
8. −$163.12
9. −$198.35
10. $551.65
11. $1,751.65
12. $1,501.65
13. $1,235.67
14. $870.67

Be Rational! (page 105)

1. a^9
2. b^8
3. y^{12}
4. z^{25}
5. 0.2
6. 0.3
7. $0.\overline{259}$
8. 0.25
9. $0.\overline{12}$
10. $0.\overline{33}$
11. 0.4

12. 0.22
13. <
14. =
15. =
16. >
17. y^4
18. z^6
19. b^0
20. a^4

Combinations (page 106)

1. 4, 5; 5, 5; 4, 6; 5, 6; 6, 6; 1/21
2. 1/5
3. 1/56
4. 1/10

Permutations (page 107)

1. 1/24
2. 1/120
3. 1/6
4. 1/24
5. 1/120

Impossible and Certain (page 108)

1. Impossible. There aren't that many dots on a nine-dot domino.
2. Certain
3. Certain
4. Impossible. You can't choose a composite from a list of primes.
5. impossible
6. certain
7. impossible
8. impossible

Calculating Chance (page 109)

1. 6/20 or 3/10
2. 3/20
3. 5/20 or 1/4
4. 1/20
5. 5/20 or 1/4
6. 15/20 or 3/4
7. 5/20 or 1/4
8. 9/20
9. 20/100 or 1/5
10. 10/100 or 1/10
11. 50/100 or 1/2
12. 12/100 or 6/50
13. 2/100 or 1/50
14. 1/4